KEEP ITS WORTH:
Solving the
Most
Common
Building Problems

This book is dedicated to the memory of Vincent A. Scaduto.

KEEP ITS WORTH:
Solving the
Most
Common
Building Problems

Joseph V. Scaduto
and
Michael J. Scaduto

TAB BOOKS Inc.
Blue Ridge Summit, PA

FIRST EDITION
SECOND PRINTING

Copyright © 1988 by TAB BOOKS Inc.
Printed in the United States of America

Library of Congress Cataloging in Publication Data

Scaduto, Joseph V.
Keep its worth: solving the most common building problems / by
Joseph V. Scaduto and Michael J. Scaduto.
p.   cm.
Includes index.
ISBN 0-8306-0361-1   ISBN 0-8306-2961-0 (pbk.)
1. Buildings—Maintenance. 2. Building inspection. I. Scaduto,
Michael J. II. Title.

TH3351.S22   1988   88-20157
643'.7—dc19   CIP

TAB BOOKS Inc. offers software for
sale. For information and a catalog,
please contact TAB Software Department,
Blue Ridge Summit, PA 17294-0850.

Questions regarding the content of this book
should be addressed to:

Reader Inquiry Branch
TAB BOOKS Inc.
Blue Ridge Summit, PA 17294-0214

Cover illustration by Mr. George Robinson.
Photographs by Ms. Stacy Scaduto.
Illustrations by Mr. Steve Bruno.

Contents

Preface

Keep Its Worth: Solving the Most Common Building Problems is a sequel to *What's It Worth? A Home Inspection & Appraisal Manual*. I wrote *What's It Worth?* in 1985 for people who are interested in buying and owning real estate. As a follow-up for those of you who have gone through the trauma of buying a home or investment property, I have written this text on building maintenance. The word "building" was substituted for the word "home" to include condominiums and investment properties. At times these two words were used interchangeably in the text.

I very much hope *Keep Its Worth* will help you avoid the many pitfalls of ownership. Certainly, the anxieties that were caused by the ordeal of purchasing property are now replaced by the uncertainties of how to "keep it up." When your budget is stretched to the limit just to make monthly mortgage payments, you surely don't want to lose any sleep worrying about how much money it is going to cost you to keep your basement from floating away, for example. In this book I answer questions about the most common building failures and problems that plague most owners. It is written in clear, simple layman's language so that any do-it-yourself owner will be able to follow the maintenance directions and, in turn, save hundreds and perhaps even thousands of dollars in repair and replacement costs. I certainly hope you enjoy reading this book as much as I have enjoyed writing it.

Acknowledgments

I am indebted to many people for their cooperation and generous assistance in preparing this book. Again, I wish to thank Ms. Rosemarie I. Zunke for her extensive editorial work. As was true with the first book, *What's It Worth*, her cooperation was again generous and helpful. Above all, I am deeply grateful to my family for all of their support. I wish to thank my wife, Tillie, for her constant support. Thanks to my daughter, Stacy, for all of her help in putting the photographs together. And a special thanks to my partner—my son, Michael, for his assistance and collaboration on this work.

Introduction

After you have bought your home, condominium, or investment property, don't think that you can just sit back and watch the grass grow greener. From the moment you first visited the building to that wonderful day you cross over the threshold to assume ownership, a silent and often invisible process of deterioration has been in progress. If the previous owners were not much on maintenance, the process began long before you came along. This is not only happening in your building, but in every other structure of any type and kind that is not properly maintained. Without careful maintenance, deterioration is inevitable.

As an owner of real estate, you should know four key terms and what their implications are: inspection, preventive maintenance, repairs, and replacements. All are keys to a "healthy" building. Don't believe for a moment that your days of inspecting are over once you pass papers on your building. As an owner, inspecting your property will now be a way of life for you. Constant vigilance for signs of deterioration and failure in your building is required if you are to avoid major costly repairs and replacements. From the casual viewing of your property as you pull up to it in your car, to the more thorough visual walk-around in the spring to see what presents winter has left you, all are part of a continued inspection process. You actually check your property every time you look at the exterior to see if it needs another coat of paint or when you go down to the basement for something. Subconsciously your mind is looking to see if everything is all right. It is this ongoing inspection process that, when combined with a good preventive maintenance program, keeps the drain on your pocketbook to a trickle.

Maintenance of a building is crucial to ensure that it stays in top condition. This book is devoted to providing you with the important maintenance information that will help you avoid major repairs and costly replacements. Each of the chapters addresses the most common building dysfunctions that afflict all kinds of properties. The format is set up to take you from the actual inspection process to a step-by-step maintenance program. To keep buildings in good shape is not only necessary for older ones but for new construction as well. It does not take long for a roof to leak, a basement to flood, or for wood members to be nibbled away by insects. Only by paying attention to the "red flags" that your property sends out to you and by giving necessary first aid can you save not only money but your precious time as well.

Procrastination on small jobs that ought to be done to keep your possession in good shape results in costly repairs and even major replacements later on. Of course, nothing lasts forever, and even with the best of maintenance and upkeep, some portions of a building will eventually need to be extensively repaired or even replaced.

Sometimes you might find yourself in a quandary about whether to repair or replace. This decision is particularly difficult to make for those who never owned property before and have little building experience to fall back on. For this reason, a life expectancy chart was prepared for you that you will find in Chapter 9. It will help you decide if it is economically feasible to just repair or if it is time for replacement. Key building parts are given with their estimated useful life expectancies. Keep in mind that many building parts, such as roofing for example, might last well beyond their anticipated life spans; some, like a hot-water tank, might not last as long as one might expect. It is always a good idea to keep all of your warranties and guarantees in a safe place. A sudden demise of a costly appliance or building component could put a deep crunch into your budget. On the other hand, if you are still under warranty, it might not cost you a dime.

There are hundreds of excellent books already on the market about all sorts of repairs (plumbing, electrical, masonry, etc.). The purpose of this book is not to rephrase what other authors have already written but rather to concentrate on the maintenance and upkeep of buildings. For your convenience, a detailed list of repair books that you might wish to refer to is given in the Appendix.

The intent of the chapters that cover wet basements, condensation, roof leaks, wood-boring insects and decay, and mechanical systems is to assist you in properly maintaining your property, as is the intent of this text. The purpose of including a chapter on energy losses is to spark new interest in energy conservation practices that the average property owner or tenant can follow and afford. The in-depth study of major health and safety problems associated with buildings was included because you have a right to know and to be informed about hazards that might be lurking in your building. A chapter dealing with contractors was added because eventually you will have to deal with them for some repairs or replacements and being forewarned is truly being forearmed.

Knowing how to deal with all the problems discussed in this book should make you feel much more like the king of your castle rather than a pauper in a hovel. After all, who wouldn't want to be the king of his own castle?

Inspecting
the Building

Your home, condo, or investment property is possibly the largest single purchase that you will ever make in your lifetime. As costs for repairs and replacements escalate, it becomes more and more important for you to understand how a careful inspection and maintenance program can save you thousands of dollars annually. By knowing what to look for and how to maintain mechanical and structural parts of your property, you will stop the clock of the deterioration process and thus prolong their life span.

The purpose of a good inspection program, whether it is followed weekly, monthly, semiannually, or annually, is to pinpoint the cause of a problem. Then, with the help of the information this book provides, you can repair the damage at a reasonable cost in the least amount of time.

If you have never inspected a building, arm yourself with a first-class text such as *What's It Worth? A Home Inspection & Appraisal Manual* (TAB BOOK #1761). This book covers every aspect of an inspection, and at the end of each chapter, you will find checklists that summarize what to look for. At the end of the text there are inspection worksheets that you can copy. These are extremely useful in "tallying" the condition of your building.

INSPECTING THE EXTERIOR

Start your inspection by examining the exterior of the structure. This chapter will give you a capsule account of what to look for during your inspection. Use the worksheet in Appendix A to make your final diagnosis as to the overall health of your building and where it needs immediate treatment.

Roof

In the spring, after the onslaught of winter snows and rains, inspect the roof thoroughly for damage that might encourage water to seep or leak into your building. If you own a one-story structure, use a ladder to get a close-hand appraisal of damages. If your property is multistoried, go about it in a safe way and use binoculars to check the roof. As you survey it, look for cracked, loose, missing, and damaged shingles or tiles. Figure 1-1 shows such damage. Also be on the lookout for damaged flashing or flashing that has pulled away, allowing water penetrations to the building. In Fig. 1-2 you can see such a potential entry point. Be sure to check the attic or areas directly under the roof for signs of leaks or damage. Mark down all your findings for future action.

Chimneys

Chimneys should be examined in the spring and before you use them in the winter. If your roof is steep and dangerous for you to climb, use your binoculars to see if your chimney has loose and missing mortar in the joints between the chimney bricks, broken or missing bricks, or if there is damaged flashing. All indicate a need for repairs and maintenance. In Fig. 1-3 you can see the loose chimney flashing and open joints that will allow water into the building. Make sure such defects are quickly repaired before they get worse. If your chimney is made of metal, be sure to look for corrosion, creosote buildup, and overall deterioration. Mark down on your worksheet all the areas that need your immediate attention.

Roof Drainage

Check your gutters at the end of the fall season and after winter to be certain that they are still doing their job. At the end of the fall, make it a point to clean out all debris

Fig. 1-1. A closeup of roof shingles in advanced stages of deterioration. This roof is ready for replacement.

2

Fig. 1-2. Old and damaged flashing is probably one of the major entry points for water. If you find such deteriorated areas, be sure to either repair or replace them.

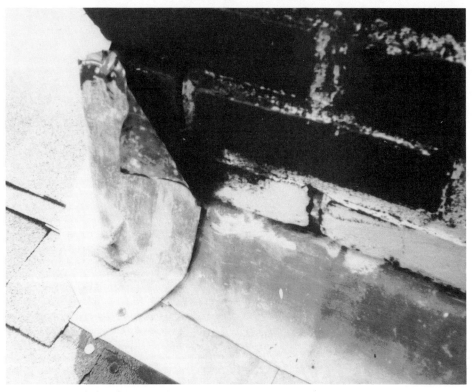

Fig. 1-3. The majority of most roof leaks are the results of defective flashing.

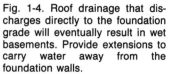

Fig. 1-4. Roof drainage that discharges directly to the foundation grade will eventually result in wet basements. Provide extensions to carry water away from the foundation walls.

such as leaves and roof granules. If your property has many trees near it, recheck once or twice more after your initial cleaning. After the winter season, gauge the position of your gutters because the weight of ice and snow could have easily made them sag. Your downspouts should always discharge water away from the foundation and not directly to them as seen in Fig. 1-4. Damage to the building and water penetrations to the basement are highly likely. If your building looks like Fig. 1-4, add extensions at the base to carry water away from the building.

Exterior Walls

Exterior walls, no matter what they are made of—wood shingles or clapboard, metal or vinyl siding, brick, or any other type of covering—should be inspected carefully every year. Does the wood look like it is starting to decay? Are there blisters in the paint? Do you see any cracked stucco or cavities in the mortar joints? If you do, try to make the necessary repairs as soon as possible. Next, take a special look at wood siding or trim that are too close to the soil or are adjacent to roof drainage runoff. As you can see in Fig. 1-5, major rot has taken place in this building. A careful inspection and maintenance program could have avoided this major repair expense. Helpful hints as how to combat rot and wood-boring insect damage are given in Chapter 5.

Fig. 1-5. The use of substandard materials, such as this fiberboard siding, will result in major deterioration.

Windows

Windows, like exterior walls, can be inspected on a yearly basis. Sometime in the fall is usually a good time to double-check the condition of putty, caulking, drip edges, and the general weather tightness of each unit. While you are at it, look for cracked glass, ripped screens, decaying trim, and peeling paint. In Fig. 1-6 you can see the beginning of some future problems. A good maintenance program would call for immediate scraping, priming, and painting of such areas to avoid the possibility of decay. Also be sure to note on your worksheets any missing hardware, storm windows, and screens. Always remember to tighten up and batten down the hatches before old man winter comes calling.

Foundation

Periodically while you are working around the outside of your property, give the foundation an eyeball examination for any defects. If the stucco finish is pulling away or cracking, or if there appears to be new cracks in your concrete foundation walls, be sure to make necessary repairs before these minor defects turn into major problems as seen in Fig. 1-7. The owner of this building did not take the time to inspect and maintain, and the results are going to be costly to repair.

Foundation Windows

Foundation windows that are low or are in well depressions are prime candidates for water flowing into your basements. A protective hood (Fig. 1-8) is a good investment.

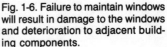

Fig. 1-6. Failure to maintain windows will result in damage to the windows and deterioration to adjacent building components.

The hood keeps water out and keeps the wood sills and window trim dry and free from rot and insect damage. Remember to annually check wood foundation windows for peeling paint and decay, and metal ones for rust.

Entrances

Every six months or so have a close look at your entrances and exits. Watch out for loose or missing railings, wood framing that is starting to show signs of decay, and metal that is beginning to rust. Are the steps safe to walk on or are they tripping or falling hazards? If you have a bulkhead entrance to your basement, make sure that it doesn't look like the one in Fig.1-9. Water leaking into your basement as well as decay and wood-boring insects are what you can expect from this poorly maintained entrance area.

Ventilation

Twice a year, scrutinize all screens for openings that might invite insects, rodents, or birds. Also check the vent openings. They must be clear of all obstructions so that air can flow in and out of attics and crawl spaces. Only if air flows freely can you prevent condensation and moisture-related problems such as in Fig. 1-10. Peeling paint and decay are the results of trapped moisture in this building. Chapter 3 tells you how to avoid such dilemmas.

Fig. 1-7. Repair open joints in foundation walls before structural damage and water penetration occur.

Fig. 1-8. If your property has deep foundation windows, protect them with hoods.

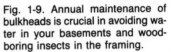
Fig. 1-9. Annual maintenance of bulkheads is crucial in avoiding water in your basements and wood-boring insects in the framing.

Garages

If you have a detached garage, treat it just like a small single-family home. Once a year check the roof, the exterior walls, windows, doors, etc., as you would with your own home. If your garage is attached to the building, then pay particular attention to garage doors and door jambs that are in close proximity to the soil. Rot and wood-boring insects favor the ends of these wood members for their upward climb to success. If you have a garage that looks anywhere near the one in Fig. 1-11, you either inherited it or you have been very lax in your maintenance program. Major settlement from decay and wood-boring insect damage makes this a prime candidate for the woodpile.

Wood-boring Insects

Because wood-boring insects, such as termites, carpenter ants, and powder-post beetles, can cause very serious damage to your property, it is a good idea to always be alert for signs of them. They are apt to invite themselves in at any time during the year. To check for the presence of such insects, probe all suspect wood members with a tool, such as a screwdriver or awl. In Fig. 1-12 you can see a prime area for both decay and wood-boring insect action. If you find such evidence, call an exterminator for a professional opinion to see if you have an active colony in your building. (Learn more about wood-boring insects in Chapter 5.)

Fig. 1-10. Failure to provide sufficient ventilation in your building will result in moisture related problems.

Driveways

Your driveway should be checked out twice a year—shortly before winter and shortly after. Cracks and otherwise deteriorated areas in concrete or asphalt can be readily repaired with today's easy-to-use products. Simple maintenance of cracks and open joints will avoid future frost heaves and excessive overall deterioration, as shown in Fig. 1-13.

Septic/Cesspool Systems

If your building does not have access to a public sewer, and you have a cesspool or a septic system, instead, then you must be alert to clues that tell you that all is not right with your system. Exceptionally lush grass over your tank or leaching field, strange odors coming up from the ground, and wet spongy sections of your yard all indicate problems with your system. To avoid costly repairs, each year pull the cover off your tank (Fig. 1-14) and inspect its interior. If it is loaded, call in your local pumping company to clean it out. At a very minimum, tanks should be pumped out every three years to avoid damaging the leaching fields. For more information about private sewer systems check Chapter 7.

Crawl Spaces

Areas under porches, room additions, and houses built over crawl spaces that cannot be easily reached are spaces much favored by wood-boring insects and decay-producing

Fig. 1-11. Treat garages like you would any building structure. Examine every component, make necessary repairs, and continue maintenance upkeep to avoid what you see in this photograph.

Fig. 1-12. Anytime that you have untreated wood in direct contact with the soil, you are tempting fate. Both decay and wood-boring insects will cause damage to such wood members.

Fig. 1-13. Annual maintenance and repairs can help you avoid major costly driveway resurfacing.

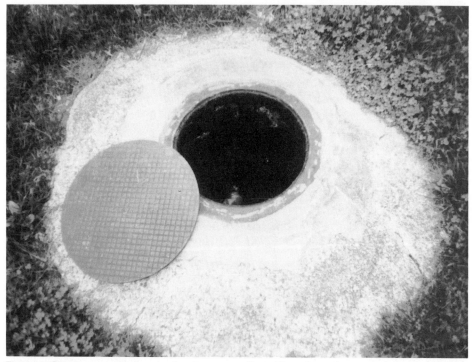

Fig. 1-14. Check your septic/cesspool systems annually and have the tanks pumped out at a minimum of every three years, more often if necessary.

Fig. 1-15. Inspect crawl spaces once a year and make corrective repairs and maintenance work to avoid structural damage.

organisms. If your crawl space does not have an access, be sure to have one put in so that you can inspect these dark and damp areas annually. If you have an access, be sure to get in there at least once a year to see what is going on.

Damage such as that seen in Fig. 1-15 could cost you several thousand dollars or more to correct. Note the sagging support posts in this figure. Major settlement and sag were evident throughout this building.

During your inspection tour, look for decay, evidence of wood-boring insects, signs of water or moisture, lack of insulation or damage to existing insulation, insufficient ventilation, missing vapor barriers, and general deterioration. If you look closely at the example in Fig. 1-15, you will find almost every possible defect that a crawl space could have. Hopefully your crawl space areas have nothing in common with the illustrated one shown here.

Fences

If your wood picket fence—should you have one—is not made of pressure-treated wood, then you have a good chance of both decay and wood-boring insect activity and damage. During the spring and summer, probe various parts of the wood fence with a tool such as a screwdriver, Pay particular attention to low areas and areas in direct contact with the soil.

Fig. 1-16. Maintain fences on an annual basis. Make sure there are no potential hazards, such as exposed metal barbs.

Metal fences also require some sprucing up as well. Once a year, sand down rust spots, prime with a metal primer, and put on two coats of a metal paint to protect it from the weather. To prevent anyone from hurting themselves on exposed barbs, as seen in Fig. 1-16, take a pair of pliers and bend these down. Not only would you not want anyone getting hurt, you also would not want a lawsuit over such a hazard.

Trees, Shrubs, Ivy, Vegetation

During your maintenance tours, be sure to record any overgrown vegetation that could cause damage to your building. Overgrown ivy (Fig. 1-17) can cause decay to wood siding and can even work its way into the building. Vines can be a highway for insects and rodents to your home, so be sure to limit or get rid of them to avoid any such negative experiences for your building. Also trim back overhanging branches to avoid clogged up gutters and damaged roofs from falling branches.

INSPECTING THE INTERIOR

The areas covered so far were for the exterior of the building. Be sure to check and inspect as I have suggested to you. The outside of the building is crucial in the preservation of the interior areas, so it is very important for you to make sure that the outside is tight and free of any defects that could affect the building. This section deals with inspecting the interior areas of a building.

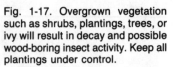
Fig. 1-17. Overgrown vegetation such as shrubs, plantings, trees, or ivy will result in decay and possible wood-boring insect activity. Keep all plantings under control.

Basement Wood-boring Insect Inspection

Wood-boring insects not only thrive on the outside but have a habit of thriving very well on the inside of a building, particularly in basements and crawl spaces that are not maintained. With this in mind, give a twice-a-year probe to all foundation sills, the ends of wood support posts, the base of partition walls, and the ends of floor joists. Make it a habit to look for sawdust droppings from carpenter ant activities, mud tunnels from termite efforts, and small pinholes in wood (as seen in Fig. 1-18), which is the work of powder-post beetles. If you find any of these signs, call in an exterminator for advice.

Foundations

Check foundations for cracks or openings (Fig. 1-19). Any opening in a foundation wall is a perfect entrance for water to flood your basement. Make sure to seal up and cement over any areas that could allow water in. For more information on wet basements and how to cure them, see Chapter 2.

Electrical

Once a year, check the grounding connection for your electric service. Figure 1-20 shows the grounding connection on the water pipe coming into your basement. If you find

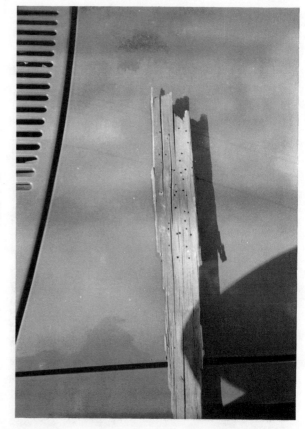

Fig. 1-18. This is a sample of wood that has been infested with powder-post beetles. The small holes in the wood are the flight exit holes that the beetles take to leave the wood after they are through feasting.

that the connection is loose, merely tighten up on the screws to make a tight bond. New codes now require a jumper cable to connect on both the street side of the water meter and the house side. Check with your local electrical inspector to see if this is required on your building. Every six months, trip all of the circuit breakers (Fig. 1-21), including the main disconnecting breaker, to make sure that they are still functional and not frozen over with rust. Chapter 7 gives you a detailed account on what to do to maintain your electrical service.

Heating and Air Conditioning

Heating and air conditioning systems should be cleaned and serviced on an annual basis by a trained serviceman. This does not mean that you get away without doing any maintenance; on the contrary, as you will see in Chapter 7, there are any number of maintenance chores that you should be doing, such as maintaining the low-water cutoff valve on a steam system (Fig. 1-22). Failure to maintain such important devices could be costly.

Plumbing

Periodically while you are in the basement working or just hiding from your spouse, examine your plumbing system for signs of corrosion, damage, or leaks. If you find any

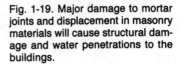

Fig. 1-19. Major damage to mortar joints and displacement in masonry materials will cause structural damage and water penetrations to the buildings.

damaged areas, such as shown in Fig. 1-23, have them repaired. On an annual basis, check all of your shutoff valves by turning them off and then back on. You wouldn't want to find out that a shutoff, particularly the main water shutoff, was defective during a flood. It is a good idea to label all shutoffs so that you know which is which in an emergency.

Domestic Hot Water

Most average hot-water tanks have an estimated useful life expectancy of anywhere from five to ten years. The older your tank is, the more frequently you should inspect it for defects. Check for signs of corrosion on the jacket and piping as well as for leaks or signs of leaks (Fig. 1-24). When you see unexplained water stains on the floor, as in Fig. 1-24, anticipate replacement before a major flood occurs.

Attics and Crawl Spaces

Every accessible attic or crawl space should be inspected once or twice a year. These areas are usually neglected or ignored altogether, even though they are the areas where problems usually occur. Condensation, leaks, decay, wood-boring insect activity and damage, deterioration in wood framing, and fungal growth are just a few things that go on without anyone knowing about it.

Fig. 1-20. During your inspection-maintenance tours, be sure that the grounding connection is secure to the main water pipe and not disconnected as seen in this photograph.

Before you inspect any attic or crawl space, be sure to put on a protective safety mask (Fig. 1-25). You never really know what harmful dust particles and fibers might be floating in these areas. Scientists are now telling us that things that were not previously considered harmful are now on an ever-growing list of suspect causes of illnesses. So the bottom line is always protect yourself.

Bathroom and Kitchens

Every six months, assess the condition of tile walls and floors for water-related damage to tile and tile joints. Gently press your hands against such wall and floor areas to see if there is much movement. Too much play means that the material behind the tile has deteriorated to such an extent that major repairs might be required. Figure 1-26 shows an open joint in a tile wall. This should be grouted and sealed to prevent water damage behind the tile. While in these areas, also check the exposed plumbing for leaks, corrosion, and damage. For a more detailed account on bathrooms and kitchens see Chapter 7.

Fireplaces and Solid Fuel Stoves

If your building has either a fireplace, a woodburning stove, or both, be sure to check each one out thoroughly at the end of the heating season and once again at the beginning.

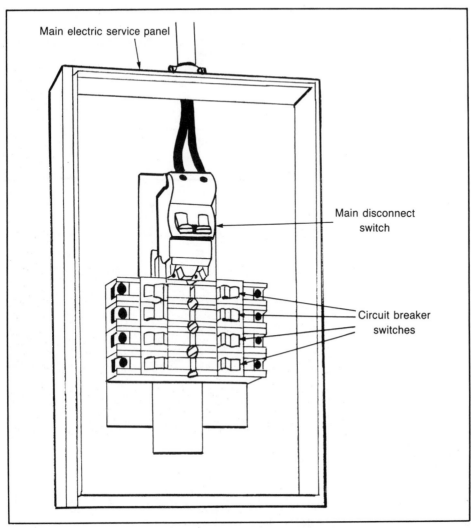

Main electric service panel

Main disconnect switch

Circuit breaker switches

Fig. 1-21. Every 6 months, trip all circuit breakers, including the main breaker. Failure to maintain these breakers could result in an inability to trip them in an emergency.

Look for excessive buildup of creosote or soot, as seen in Fig. 1-27. Check for damaged brick or flue pipes. Woodburning stoves and fireplaces that are used extensively should be professionally cleaned and inspected on an annual basis. Chapter 8 will give you helpful hints on how to avoid potential hazards when using these supplementary heating devices.

Rooms

Once a year, check electrical outlets in your building. You want to be sure that all are in good working condition and functional. If you find loose mounted outlets, such as the one in Fig. 1-28, be sure to have an electrician correct this potential hazard.

In the fall, check all windows to make sure that sash cords are not ready to give up and allow the window to come crashing down. Figure 1-29 shows a sash cord that is

Fig. 1-22. If you have a steam heating system, follow the manufacturer's directions to drain the low-water cutoff valve. Note the plug cap in the discharge pipe. Obviously the owner is not performing this important maintenance chore.

frayed and should be replaced before an accident occurs. During the cold, drafty winter days and nights, check for cold drafts around windows and doors. Caulk and weatherstrip all such energy gaps. Chapter 6 reviews some common sense energy conservation.

SUMMARY

Because it would be impossible to do all of the inspection and maintenance chores in 1 day or in a short span of time, you should try to devise some plan of attack. Chapter 9 gives you a variety of schedules that you can either use or modify to meet your individual needs. Don't worry if you forget to do some chore—your building will not fall apart on you overnight. Just catch up on the work sometime later. Never make your inspection and maintenance program a dreaded chore that you will try to avoid doing. Rather, make it a natural part of your ownership. Think about how much more valuable your property will be after you give the exterior a coat of paint or brighten up a room with a face-lift.

The checklist at the end of this chapter should help you remember what to check out. The worksheets provided in Appendix A will help you keep records of what you have done and what you should be doing. As you read Chapters 2 through 9, you will find many useful maintenance and repair tips that will help you maintain and keep the worth of your investment.

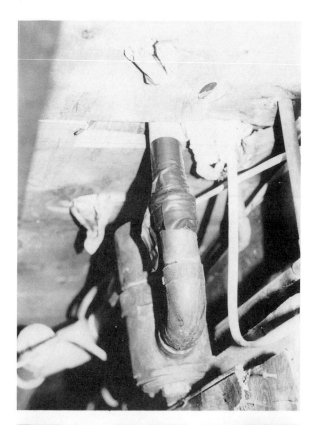

Fig. 1-23. Instead of make-shift repairs, as seen in this photograph, make correct repairs or replacements to avoid the potential of water damage to your property.

Fig. 1-24. Monitor hot-water tanks on a regular basis. Look for signs of deterioration and replace a tank at the first sign of failure.

Fig. 1-25. Always remember to protect yourself by wearing a safety mask whenever you enter attics or crawl spaces.

Fig. 1-26. Maintain joints in tile walls and floor areas by grouting and caulking periodically.

Fig. 1-27. Never allow creosote to build up to the extent that you can take handfuls out, as seen in this photograph.

Fig. 1-28. Always be on the lookout for loose or defective outlets and electrical switches. Make necessary repairs immediately to avoid the possibility of shock or injury.

Fig. 1-29. Once a year inspect for and make repairs to windows and their component parts. Replace frayed sash cords, repair deteriorated putty, and caulk and seal open joints.

A word of *caution* should be interjected here. Under no circumstances should you put yourself in any danger during your inspection and maintenance program. If any project appears to be hazardous or dangerous, by all means hire a professional to do the job.

CHECKPOINTS

Exterior

- Examine roof areas for signs of deterioration and damage.
- Check chimneys for damaged brick or rusted metal.
- Look for lifted flashing on chimneys and roof areas.
- Inspect roof drainage for decay or damage.
- Make sure all downspouts extend several feet from foundations.
- Note damaged siding and trim.
- Inspect for air leaks around windows and doors.
- See if decay has taken place on wood members.
- Check all entrances to see if they are safe.
- Double check screens on attics and crawl spaces.
- Inspect garages like you would a building.
- Be on the lookout for evidence of wood-boring insects.
- Mark down on your maintenance sheet areas of driveways and walkways that need repairs.

- Pull the cover off your septic or cesspool tank and inspect the interior.
- Get into crawl spaces that are accessible and look for damage.
- Check fencing to see if it needs maintenance or repairs.
- Trim back all vegetation too close to your building.
- Use the schedules given in the text or make up your own to establish time frames for inspections and maintenance work.

Interior

- Examine basement framing for decay and wood-boring insect damage.
- Inspect for signs of water leakage or seepage into the basement.
- Check for signs of condensation.
- Inspect the electrical-mechanical system for defects.
- Survey attics and crawl spaces for deficiencies.
- Note fireplace and wood burning stove defects.
- Be on the lookout for water stains on ceiling and walls.
- Be alert to hazards in the building.
- Caulk and seal open joints in floor and wall tiles.
- Monitor plumbing for leaks or damage.
- Examine electrical receptacles for damage.
- Make sure all outlets are working.
- Plug up any air leaks in the building.
- Develop and follow a schedule for regular interior inspections and maintenance.

2

Wet Basements

Of all the annoying problems you as a building owner might encounter, a wet basement is undoubtedly one of the most disconcerting; disconcerting in that a basement, unlike a roof, cannot always be waterproofed completely and satisfactorily. If you are unlucky enough to have a building in a high water table area, you might be among the many who have recurring wet basements.

In this chapter I will try to answer some of the most common causes for wet basements. You will learn about foundation defects, faulty roof drainage, improper lot and foundation grading, and poor design of foundation windows and entrances, as well as about natural phenomena such as high water tables, underground springs, and soil conditions that lead to wet basements. Remedies and professional tips to alleviate and possibly eliminate wet basements will be covered.

Statistics tell us that, at one time or another, most homes will suffer damage in the basement because of water. If you don't want to be the little Dutch boy with his finger in the dike, you should plan a course of action and eliminate the possibility of your finding an unplanned swimming pool in your basement after a heavy rainfall. There are many preventive measures that you can take to eliminate water in your basement. As seen in Table 2-1, some are relatively simple while others require professional help. Faulty roof drainage, for example, is a simple matter to correct, whereas doing something about a high water table will be an expensive undertaking requiring the services of a professional contractor.

Table 2-1. Wet Basement Repair Steps.

Simple Steps

✔ Improve or add roof drainage to your building.
✔ Improve foundation grading.
✔ Caulk and seal any open joints in bulkheads, walkways, and windows.
✔ Provide hood covers for foundation window wells.
✔ Limit or remove foundation plantings.

More Involved Steps

✔ Seal cracks in foundations (point up joints).
✔ Repair deterioration in stucco walls.
✔ Waterproof interior foundation walls.
✔ Provide catch basins.
✔ Install sump pumps and dry wells.

Professional Help Needed

✔ Provide exterior perimeter drainage system.
✔ Waterproof the exterior of the foundation.
✔ Provide interior buried drainage system.
✔ Provide interior surface drainage system.

CAUSES OF WET BASEMENTS

You might have a wet basement because of condensation, seepage, or leakage. *Condensation* is fully covered in Chapter 3. The difference between seepage and leakage is one of degrees. In the case of *seepage*, water seeps in through joints slowly; in the case of *leaks*, water can flow rapidly through holes and cracks. Table 2-2 reviews the causes and remedies for each of the three causes. All of these causes will result in damage to your building, so it is important for you to review what signs to look for and the solutions to correct them, as explained in Table 2-2.

Roof Drainage

Let's look at the easy jobs first. Probably the most common cause of wet basements is faulty roof drainage. Downspouts that discharge directly to foundation walls are a major culprit. A simple solution is to add extensions (Fig. 2-1). When such extensions are installed, the roof runoff is discharged several feet from the foundation walls.

Lack of gutters on a building can be another cause of wet basements. A building such as the one in Fig. 2-2 will have its rain and melting snow dropping right down to the foundation with a good chance of its working into the basement. In addition, water pouring off a roof will, in time, erode the top soil along the foundation, leaving a barren look. The splash from such water could affect the low wood members and rot might take over in time. If your roof has an overhang less than two feet, consider installing gutters and downspouts.

Table 2-2. Condensation—Seepage—Leakage Identification.

Signs of Trouble	Causes	Solutions/Remedies
Condensation		
Damp walls	Humid air in the basement (see Chapter 3 for a complete explanation).	Adequate ventilation is the basic cure for condensation.
Mildew and mold Sweating pipes and window surfaces Corrosion on metal objects Musty odors		
Seepage		
Damp spots on foundation Efflorescence on walls Oxidation on metal Decay and deterioration	Small holes or fine cracks in the foundation walls as well as deterioration in the foundation coating.	Improve roof drains. Repair cracks with hydraulic cement. Waterproof the foundation walls.
Leakage		
White crust on floor tiles and joints Raised and buckled floors Rotting framing members Cracked walls Water stains Musty odors	Hydraulic pressure from ground water, high water table, damaged floor and walls.	Exterior waterproofing, install exterior/interior drains, add sump pump, make repairs to foundation and floor areas.

If you presently have gutters, be sure to clean and repair them every 6 months. Figure 2-3 shows gutters that have been badly neglected. Gutters loaded with leaves and debris will cause rot and draw wood-boring insects. The owner of this building has not done any maintenance in a long time, and the building is showing the signs of it.

Foundation Grading

Proper foundation grading is crucial in your program for keeping a dry basement. Figure 2-4 shows a good example of how your foundation soil should not slope. In this instance, water will collect and seep into the basement. The simple solution would be to fill the hole and slope the soil so that a gradual slope forms away from the foundation.

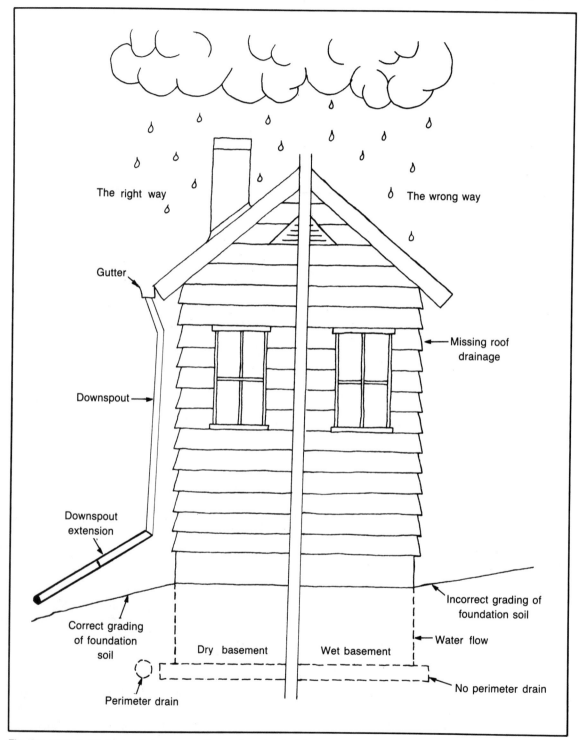

The right way

The wrong way

Gutter

Missing roof drainage

Downspout

Downspout extension

Incorrect grading of foundation soil

Water flow

Correct grading of foundation soil

Dry basement

Wet basement

Perimeter drain

No perimeter drain

Fig. 2-1. How to discharge drainage runoff away from the building.

Fig. 2-2. Missing gutters will result in water in basements, erosion of foundation soil, and damage to low wood siding and trim.

Foundation Windows

Foundation windows, particularly low ones such as the one in Fig. 2-5, act as catch basins for water (rain or snow) and allow seepage into the basement. By installing hood covers over these areas, you can eliminate the possibility of a wet basement. Hoods can be purchased at most lumber or supply yards for a nominal cost.

Bulkheads

Wood or metal bulkheads are notorious for allowing rain or melting snow to enter basements. During a heavy rainfall, check your bulkhead to see if there are any leaks. If you find areas where water is coming in, be sure to mark it at the location with a pencil and later caulk and seal the defective opening. If you find rotted wood or rusted metal, be sure to make the necessary repairs to avoid future wettings.

Tree Roots

Roots of some water-seeking trees can raise havoc with foundation walls. Weeping willows, poplars, and birches all tend to send out strong root systems, and if your building is constructed of a foundation with joints such as stone or hollow block, you could be on the receiving end of trouble.

Fig. 2-3. Failure to maintain gutter systems can cause water to back up into the building. Clean out gutters at least twice a year, more often if necessary.

Fig. 2-4. Depressions and low spots along foundation walls will result in seepage into the basement areas. Regrade all low spots so that water will flow away and not towards your building.

Fig. 2-5. Deep well foundation windows without protective hood covers will end up with what you see here—decay, wood-boring insect activity, and water in your basement.

Figure 2-6 shows a root system that managed to get through a stone foundation wall. When these roots die and rot, they leave natural channels for water to enter the building. A good policy would be to avoid planting any trees too close to your building, particularly ones that will cause these problems. If you find the remains of tree roots in your basement, be sure to dig them out and cement up any openings in the foundation walls.

Foundation Plantings

Limit flower beds and any type of foundation plantings or eliminate them entirely from the exterior perimeter of the foundation. Overgrown vegetation tends to hold moisture, and frequent waterings by owners (Fig. 2-7) can be a source of unwanted water in the basement. If you are a flower lover, try to plant them away from the building, if at all possible, or use species such as hostas that require little watering.

Ledge

If your building foundation walls or floor are built directly on or over ledge or rock formations, there is a good chance of continuous seepage into your basement. In Fig. 2-8 you can see a ledge outcropping in a basement with concrete flooring poured around it. It is difficult if not impossible to seal such formations, particularly in high water table areas. Your best bet for a basement such as this is to install a sump pump.

Low Spots

The base of a declining driveway or the bottom of a set of stairs leading into a basement are frequent locations where water collects and works its way into the

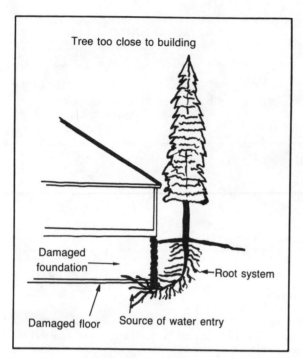

Tree too close to building

Damaged foundation →

← Root system

Damaged floor

Source of water entry

Fig. 2-6. The roots of trees planted too close to the foundation can damage foundation walls as well as cause water to penetrate the building.

Fig. 2-7. Frequent waterings to foundation plantings can result in damp basements.

Fig. 2-8. Foundations that are built on and over ledge formations will usually be the recipient of unwanted water. By maintaining foundation grading and roof drainage runoff, you might be able to curb the amount of water seepage.

basement. A catch basin is one way to redirect the flow of water away from the building. A catch basin, such as the one in Fig. 2-9, is just a large hole in the ground filled with crushed stone and covered with a grille. It can also have a pipe to carry the water away. If you have such water entrance areas, you should consider having a catch basin installed.

FOUNDATIONS

Most foundations are subject to water pressure both from lateral movement through the foundation and upward movement through the basement floor near the foundation. Depending upon the type of foundation and its condition, the results of water penetrations could be slight or major. Let's take a look at the various kinds of foundations, the types of damages you might encounter, and some cost-effective ways to repair such damages.

Brick Foundations

If you look closely at the photograph in Fig. 2-10, you will see how water can find its way into this particular building. Open and deteriorated brick joints form a perfect waterway, particularly those in the soil. The open joints around the oil fill and vent pipes also are helping out. Note the screwdriver to the right of the oil pipe. Some of these openings are large enough for small rodents to squeeze through.

Open joints in brick, stone, or concrete block can easily be repaired (pointed up). Figures 2-11 through 2-14 show the four basic steps for pointing up mortar joints. Before

Fig. 2-9. Catch basins are a good way to capture flowing water in low spots to either divert or contain the water.

Fig. 2-10. Be sure to point up defective joints and openings in foundation walls.

34

Fig. 2-11. Scratch out defective joints with a tool such as a screwdriver to remove loose and deteriorated mortar.

Fig. 2-12. Use a hose or a moist brush to wash away dirt and loose particles from the joints you are repairing.

Fig. 2-13. Pack plenty of new mortar into the open joints. When you think that you can't apply any more, try forcing some more in.

Fig. 2-14. After the joints are full, wait a few minutes and then strike each joint. Use firm pressing motions, first in the horizontal joints, then finish off with the vertical ones. Your results should be a fully packed and tight joint.

you apply the mortar, be sure to first thoroughly clean out all the joints that need repairs. Take a screwdriver and scrape out all the loose mortar (Fig. 2-11). After all the joints have been raked, wet them down with a hose (Fig. 2-12). This will get rid of leftover mortar and loose particles. Water not only cleans out the joints, it also dampens them so that the newly applied mortar can form a firm bond with the old brick. Mix your cement according to the manufacturer's directions and then apply it as shown in Fig. 2-13. Be sure to really pack it in nice and tight. When you think that the joint won't take any more mortar, force in some more. Once the joint is full, allow it to set for approximately 15 minutes. Then with a striking tool (Fig. 2-14), strike the joints in such a way to make a flush joint. Gently draw the tool across the joints, first horizontally and then vertically. This will give you a watertight joint.

Stone Foundations

Stone foundations, like brick ones, are damaged in their mortar joints first. Thus, they can be repaired in the very same way. Figure 2-15 shows a badly damaged stone foundation. Follow the steps outlined for repairing brick foundations, and you should end up with a tight foundation.

Concrete Block Foundations

This third type of foundation can give you as much trouble as the other two (brick and stone) because its joints are also built with mortar that will, in time, succumb to

Fig. 2-15. You will need more mortar to fill in the open joints and gaps in deteriorated stone foundation walls, but the principle is the same. Use chinks of small stones to fill in larger openings, then cement them in place.

the elements. Again, follow the same procedure for block walls that are outlined for pointing up brick and stone walls.

Stucco Finished Foundations

Figure 2-16 shows a foundation with damaged stucco. You can find stucco finish on any type of foundation. Usually it will be on poured concrete or on concrete block walls. When it reaches the condition seen in Fig. 2-16, you might want to call in a mason to do the work. Stucco finish repair work, unless done correctly, will, in time, break down and cause additional damage to adjoining areas. If during your inspection maintenance chores you find small sections of damaged stucco, you certainly can make some minor repairs to avoid the future possibility of major repairs. In making your repairs, be sure to tap out all the loose sections of stucco, wet down the areas to be repaired, and apply a recommended masonry material to the damaged areas. The application is usually applied in uneven, circular motions. Follow whatever pattern that already exists on your walls.

Poured Concrete Foundations

Probably the most solid of all foundations is the poured concrete foundation. Even this kind of foundation, however, will accede to water penetrations, particularly at cracks

Fig. 2-16. Be sure to tap out loose sections of stucco so that the areas all are flush to each other. Recement and seal up any such damaged areas before they increase in size.

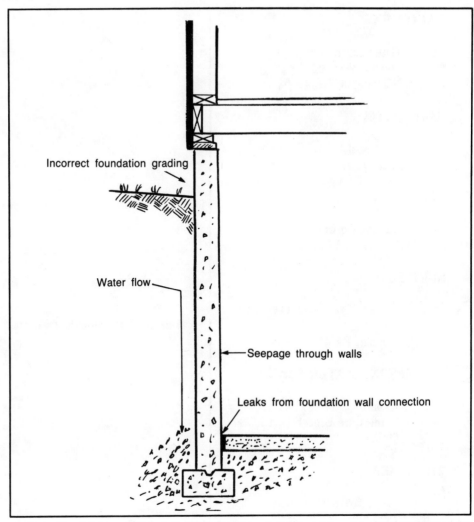

Incorrect foundation grading

Water flow

Seepage through walls

Leaks from foundation wall connection

Fig. 2-17. Check floor-wall joints periodically for signs of seepage. Ask your local hardware store for a good waterproofing material that you can apply. Follow the manufacturer's directions.

in the concrete and at the joint between the foundation wall and the floor of the basement. Figure 2-17 illustrates this point of entry in the floor wall connection.

Minor cracks, such as those caused by initial settlement, are usually easily sealed up with any one of the many fine products made for waterproofing. A listing of these products and the names of the manufacturers can be found in Table 2-3. As with any product, be sure to follow—to the letter—the manufacturer's directions as to how to apply and maintain their sealing compound.

Cracks that have developed recently and those that measure more than ¼ inch might have to be checked out by a professional to determine why the foundation is cracking and what to do about it. Either a structural engineer or a professional building inspector should be able to help you out. Your local Yellow Pages will most likely be the best source for the names of such professionals.

GACOFLEX

Gates Engineering
100 S. West St.
Wilmington, DE 19801

KOOL PATCH

Kool Seal
Clearwater, FL 33520

LEAK STOP

Bil Dry-On Corp.
Norfolk, VA 23509

SURE-SEAL

Carlisle Tire and Rubber Co.
Box 99
Carlisle, PA 17013

THOMPSONS WATER SEAL

Richardson-Vicks, Inc.
Home Care Branch Division
Memphis, TN 38117

THOROSEAL

Thoro Systems Products
7800 NW 38th St.
Miami, FL 33166

WARM-N-DRY

Owens-Corning
Protective Coatings Division
Toledo, OH 43659

Table 2-3. Waterproofing Products.

Poured concrete will also leak at the metal tie rods in the foundation walls (Fig. 2-18). These metal rods in the foundation act as a sieve and draw water into the basement. You can cut them off flush with the foundation and apply an asphaltic cement sealer. If this doesn't work, then you might have to consider digging down on the outside of the foundation to find these metal rods and seal them up at their source. Before you start digging though, consult a professional to see if that would be a wise thing to do.

40

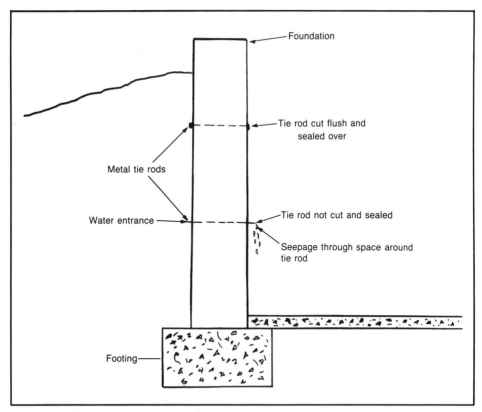

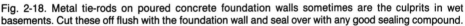

Fig. 2-18. Metal tie-rods on poured concrete foundation walls sometimes are the culprits in wet basements. Cut these off flush with the foundation wall and seal over with any good sealing compound.

BUILDINGS ON SLOPES OF HILLS

If you have a building on the slope of a hill or at the base of one, it is highly likely that you will at some time receive unwanted water into your basement areas. If you never have, you are lucky. For those of you that have such a problem, look at Fig. 2-19. This is a simple horseshoe ditch that is built to intercept rushing waters and designed to carry them away from your property. It is not something that the average owner would think about doing, but rather something that a professional with the right construction equipment could easily do. It is costly to have built, but in most instances, if done right, will solve the problem. As with any project involving large costs, be sure to get several opinions and cost estimates before committing yourself.

BUILDINGS LOCATED IN HIGH WATER TABLES

If your building is located in a flat high water-table area, you will need something quite different than what is described in Fig. 2-19. You don't have the natural slope to help you drain the water away from your property. What you need is a combination of buried drainage with a sump pump to eject the water some distance from your property. Figure 2-20 shows a typical exterior buried drainage system. This type of system is usually put in when the building is being constructed. If you have an older home without an exterior

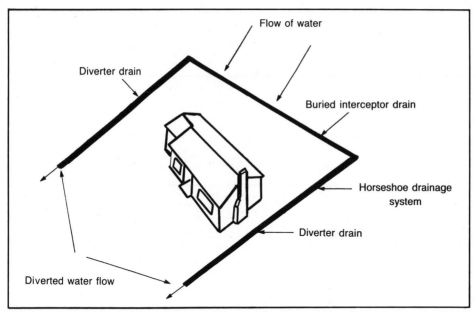

Fig. 2-19. A horseshoe exterior ditch around your building is something that a contractor will have to do. It is an expensive proposition and should not be considered until other options are exhausted.

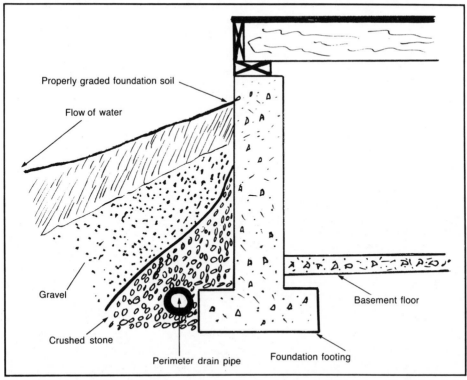

Fig. 2-20. Buried perimeter drainage systems are common on newer construction, particularly in areas with high water tables. If you are having a home built, consider this option for keeping your basement dry.

Fig. 2-21. This older form of interior perimeter drainage system is simply an open trench that leads water to a sump pump. The water is discharged out of the building, usually several feet away.

drainage system, you can have one like Fig. 2-21 installed in the basement. Both types of buried systems work on the same principle. They use perforated pipe laid on crushed stone and connected to a sump pump which is activated by rising water. In most cases, this usually solves the problem.

A relatively new twist to controlling water in basements is the installation of surface-mounted interior basement drains (Fig. 2-22). These systems consist of an interior drain system that is cemented along the base of the inside foundation walls of the basement. Holes drilled into the foundation base allow water to run into the drain system that is hooked up to a sump pump. The sump pump discharges the water several feet from the building (similar to the other systems). The advantage of these systems is that you don't have to dig up the perimeter of the basement foundation floor areas. The disadvantage is that it is only practical with foundations that can be easily drilled into, such as hollow core block foundations. For information about these systems, write to:

Channel Drain
401 Olive Street
Findlay, OH 45840

Beaver Drain Systems
1375 Laurel Ave.
St. Paul, MN 55104

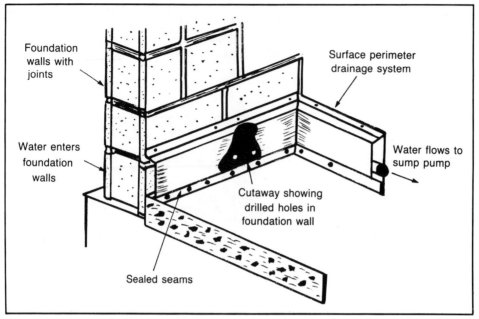

Foundation
walls with
joints

Surface perimeter
drainage system

Water enters
foundation
walls

Water flows to
sump pump

Cutaway showing
drilled holes in
foundation wall

Sealed seams

Fig. 2-22. This example of a newer interior surface drainage system is usually good for walls with existing joints such as stone, brick, or block.

If your water problem is minor or isolated to a particular part of your basement, the cure can be as simple as installing a sump pump that can handle times when the ground becomes saturated with melting snow or heavy rains. Figure 2-23 shows a typical simple sump pump installation. The corner where the water collects will be the sump pump site. Follow the manufacturer's recommendations for the sump hole and the electrical hookup. Be sure to have the discharge line end several feet from the foundation, otherwise you will be just recycling the same water in and out of your basement.

DRY WELLS

A dry well is nothing more than a hole in the ground that is used to absorb the discharge of a sump pump (or roof drainage system). If it is built correctly, it should have no trouble handling your sump pump discharges. To build one, just dig a large hole several feet from your foundation. Fill the hole with stone, broken bricks, and almost anything that will not absorb water (I prefer to leave the holes empty). If you want to get fancy, you can put a salvaged, perforated steel drum in the hole and connect your sump pump discharge pipe directly to it. Regardless of the type of dry well that you put in, be sure that it is at least 10 feet from the foundation. This will prevent seepage from the well back into the basement. Figure 2-24 shows what a completed dry well should look like.

DAMPPROOFING

Dampproofing or waterproofing is usually done at the time the building is constructed. An asphalt foundation coating is painted on the exterior of the foundation to prevent seep-

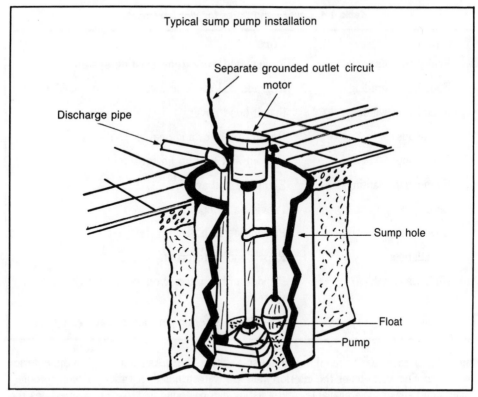

Fig. 2-23. A sump pump is a relatively simple way to keep your basement dry. In most instances the installation of one pump in the lowest part of your basement will do the trick.

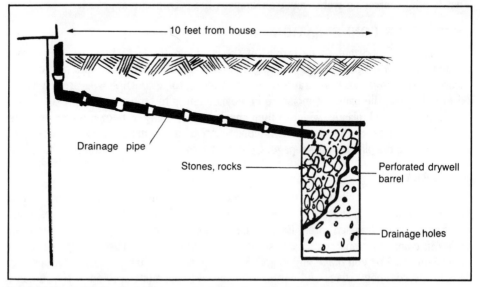

Fig. 2-24. Dry wells are discharge areas that are built and installed to absorb water from sump pumps and roof drainage systems. Keep them at least 10 feet from your foundation walls.

Table 2-4. Causes and Cures for Wet Basements.

Cause	Cure
Faulty roof drainage	Repair, add, or extend existing system.
Foundation grading	Regrade soil to slope away from the building
Foundation windows	Install hood covers
Bulkheads	Repair, replace, and seal all joints.
Tree roots	Dig out and seal up holes.
Foundation plantings	Eliminate and avoid.
Ledge	Provide a sump pump.
Low spots	Provide a catch basin.
Foundations	Repair and waterproof.
High water table	Provide either interior or exterior drainage system.
Slopes	Provide an exterior perimeter drainage system.

age. Experience has proven that this doesn't always work. As a follow-up to the exterior waterproofing, sometimes the interior walls are waterproofed. A variety of waterproofing paints and sealers are available, and a listing can be found in Table 2-3. Sometimes the results are not satisfactory, particularly if there is negative water pressure forcing its way through the foundation walls. This pressure can push the material off the surface. That is why controlling water must encompass more than one approach.

SEEPAGE OR CONDENSATION

It is very important to determine if the problem that you have is seepage or condensation. *Seepage* (water trying to get in from the outside) needs different control measures than *condensation* (trapped moisture trying to get out). A well-proven method to determine the cause of moisture in a basement is to tape a section of the inside foundation wall with plastic, which is then checked after a day or two. If water droplets are on the outside of the plastic, it is condensation; if the droplets are between the foundation and the plastic, it is seepage. Controls for condensation are found in Chapter 3.

SUMMARY

Most often, a combination of the outlined causes contributes to the fact that a basement is wet. It is therefore necessary to plan carefully whether to make exterior repairs or interior drainage cures. Each of the solutions and remedies given in this chapter varies in complexity and costs, and some will require the expertise of a professional contractor. Before you make any major improvements, be sure to consult Table 2-4 to review the common causes and cures for most wet basement problems. If you can't fathom why you have a "wet" basement after having read and re-read all of this material, call in a professional to solve the mystery.

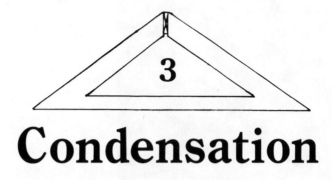

Condensation

Condensation, which is caused by humid air in attics, crawl spaces, exterior wall cavities, and basements, is sometimes difficult to distinguish from other sources of moisture. Determining the cause of the moisture is one way of telling whether it is condensation, seepage, or leaks. After the cause is determined, the next step is finding a solution. This chapter is devoted to helping you identify condensation, determine its source of origin, and hopefully provide you with solutions to remedy the problem.

Condensation can occur anywhere in a building, but perhaps because of poor airflow in some areas of the building, it is more likely to happen in attics, crawl spaces, interior portions of outside walls, and basement areas. What happens in these areas is that trapped vapors condense into water. Vapors build up and saturate the air. The air cannot hold the moisture, and condensation occurs such as can be seen in Fig. 3-1. This will result in rotting wood framing, rusting metal, and general deterioration throughout the building but particularly in the noted areas.

SYMPTOMS OF CONDENSATION

How will you know if your building has problems associated with having too much vapor in the air? An easy way would be to purchase a relative humidity gauge that could be mounted on your desk or on a wall. The best method would be to have more than one that can be placed in strategic locations, such as your basement and attic.

Other ways to tell is by just using your senses: smelling, seeing, and touching. Is there a musty odor in your building? Does the basement feel damp when you are there? Foggy windows, sweating pipes, and frost or water beads on surfaces are also signs

Fig. 3-1. Droplets of water from condensation. In addition to this visible condensation, many buildings have severe concealed condensation problems that cause extensive damage and deterioration.

of condensation. If freshly painted surfaces begin to peel after a relatively short time, you might have inadequate ventilation that would result in condensation. Table 3-1 lists the most common symptom of condensation. Use it to help diagnose moisture problems in your building.

One way of telling if the problem is condensation in a basement is by attaching a small mirror flat against a portion of the basement wall that has a damp spot. If after 24 hours you see moisture droplets on the surface of the mirror, the problem is probably condensation. Be sure to check the other side of the mirror as well. If you find dampness there, the problem could be a combination of condensation and seepage.

CAUSES OF CONDENSATION

Table 3-2 lists the major causes of and cures for condensation. Be sure to review the causes and examine the cures carefully. Most of your condensation headaches probably can be corrected by the prudent application of the suggested cures in this table.

Minimum Ventilation

If your building has no ventilation as seen in Fig. 3-2, or minimal ventilation as seen in the gable vent in Fig. 3-3, you should seriously consider having a roofer or carpenter install additional vents. The combination seen in Fig. 3-4 is likely the best method to

Table 3-1. Symptoms of Condensation.

✓ Musty odors
✓ Sweating pipes
✓ Mildew on walls and ceilings
✓ Efflorescence on walls
✓ Frost or ice on window panes
✓ Icing on exposed nails in attics
✓ Frost on attic or crawl-space sheathing
✓ Discoloration on wood framing
✓ Wet or soggy insulation
✓ Damp walls, floors, and ceilings
✓ Moist or wet rugs
✓ Mold on stored clothing in closets
✓ Peeling paint—either exterior or interior
✓ Decaying wood
✓ Deteriorated plaster
✓ Cracks in ceilings or walls
✓ Droplets of water on walls, ceilings, or floors
✓ Bowed, cupped, or swollen wood-base framing materials
✓ Dampness in the air
✓ Unexplained water stains
✓ Moist or wet exposed soil in crawl spaces
✓ Corroded metal surfaces
✓ Rusting pipes
✓ Unexplained puddles of water

ventilate a building. Minimal ventilation is the number one cause of condensation woes in most buildings. Keep in mind that you really cannot over ventilate a well-insulated building. So, don't forget—ventilate.

If you do have vents in your building, check them once a year to make sure that they are not blocked off with dirt, insulation, or debris. Sometimes soffit vents (Fig. 3-5) become victim to a painters paintbrush. Make sure that this doesn't happen in your building. Figure 3-6 illustrates a variety of vents that can be used in ventilating your building.

Vapor Barriers

Lack of vapor barriers or vapor barriers installed incorrectly can add to your condensation problems. As illustrated in Fig. 3-7, moisture can get into wall cavities and condense on vapor barriers that were incorrectly positioned. In this case, the vapor barrier should have been facing the interior wall rather than the exterior wall. As you can see, condensation occurs when the moisture reaches the cold exterior wall. The results of this over a period of time could be major decay in the building framing as well as damaged insulation.

Table 3-2. Condensation: Causes and Cures.

Causes	Cures
Minimum building ventilation	Increase ventilation by installing gable, soffit, ridge, roof, or a combination of vents.
Obstructed or blocked vents	Reopen.
No ventilation	Add sufficient vents.
No vapor barriers	Add if possible. If not possible, consider a vapor barrier paint.
Insufficient vapor barriers	Add more or use a vapor barrier paint.
Vapor barriers incorrectly installed or damaged	Correct or repair.
Insulation installed without a vapor barrier	Paint ceilings of top floors and all exterior walls with a vapor barrier paint.
Exposed soil in basement or crawl spaces	Add a ground cover of 6-mil polyethelene.
Damaged or missing sections of ground cover	Repair or replace.
Nonbreathing siding such as vinyl or aluminum	Increase building ventilation
Improperly vented bathrooms and kitchen exhaust vents	Vent to exterior.
Improperly vented dryers	Vent to exterior.
Improperly vented heating devices such as space heaters	Have serviced and vented to the exterior.
Drying firewood indoors	Dry outdoors
Poor drafts in chimney	Increase chimney height. Make repairs to chimney.
Drying clothes indoors	Use a dryer or hang clothes outdoors.
Energy efficient tight buildings	Consider an air-to-air heat exchanger. Use dehumidification methods.
Closed off, unheated spaces	Either ventilate or heat.
Overgrown trees, bushes, and vegetation too close to building	Trim back sufficiently to let the building breathe
Bathrooms and kitchen	Provide exhaust vents to the outdoors.
Use of humidifiers	Limit or discontinue use.

Table 3-2. Continued.

Causes	Cures
Poorly designed crawl spaces	Provide an access for inspection. Add ground covers to exposed soil, ventilation units, vapor barriers as well as insulation.
Wet basements or crawl spaces because of poor grading	Regrade foundation soil.
Wet basements or crawl spaces caused by roof runoff	Provide gutters and downspouts that extend several feet from foundation walls.
Loose house windows	Tighten up, winterize.
Missing storm windows	Add.
Storm windows without weep holes	Drill weep holes—one at each end of the window sill where the storm window rests.
Inefficient and dirty heating systems	Clean, service, and tune-up.
Use of a humidifier in a warm-air heating system	Disconnect and eliminate from system.

Fig. 3-2. Having no ventilation in a building is like wrapping your entire body in a plastic sandwich bag. The problem is compounded here by the vinyl siding. Concealed and visible condensation is rampant in this building.

Figure 3-8 shows a classic example of a vapor barrier installed upside down and damaged. Keep in mind that for vapor barriers to be effective they must face towards the heated side of the building and not as seen in this photograph, facing away from the heated side. Moisture that gets trapped between the vapor barrier and a ceiling or wall will cause decay and deterioration. If your building has anything that looks like this photograph, be sure to follow the suggested cures in Table 3-2.

Fig. 3-3. By having minimal ventilation in a building, you create the same situations for condensation that no ventilation causes. New technology is not always an improvement as seen in this newer type gable vent. The surface areas for air movement are restricted by the design of this vent.

Exposed soil in a crawl space (Fig. 3-9) is a key area of moisture transfer. Note the deteriorated insulation. Crawl spaces such as this one should have a proper ground cover (Fig. 3-10). Polyethylene plastic sheets, overlapping each other by several inches and held in place by bricks, is the answer for moisture transfer from exposed soil.

Vinyl/Aluminum Siding

Certain types of siding materials can be tied into the condensation cycle. Vinyl and aluminum siding are two types of siding materials that tend to prevent a building from breathing. Figure 3-11 shows a building wrapped in vinyl siding. Note the damaged sections under the window. Moisture buildup in the walls could cause enough warping in the wall framing to cause the vinyl to snap out of its interlocking channels. An inspection of this building disclosed exactly that cause. If you have a building that is sided with either vinyl or aluminum, make sure that the building has ample ventilation. You certainly don't want your building looking like this one.

Kitchen and Bathroom Vents

Improperly vented kitchen and bathroom vents are a leading cause of excess moisture accumulating in a building. If you have bathrooms with ceiling vents (Fig. 3-12), make sure that these vents do not vent directly into attics or crawl spaces (Fig. 3-13). It is surprising how many building owners allow moisture from bathroom and kitchen vents

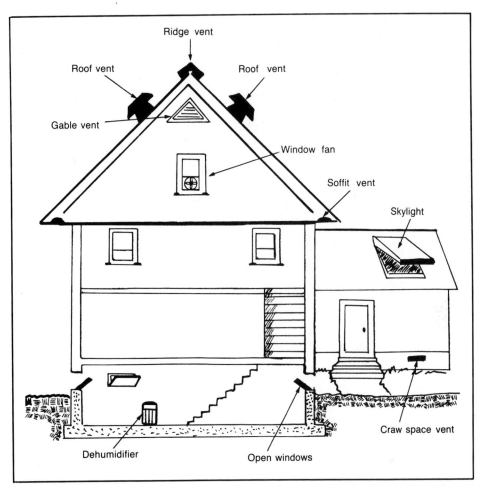

Fig. 3-4. The best way to ventilate a building is by providing a combination of vents that will react with each other to create positive flow patterns of air movement into and out of a building.

to discharge directly to these enclosed areas. The results usually are major decay and damage.

Dryer Vents

Clothes dryers must be vented to the exterior of a building to avoid massive moisture buildup in the building. Do not vent your clothes dryer into attic or crawl space areas as seen in Fig. 3-14. And don't use any of the heat-reclaimer units like the one in Fig. 3-15 to provide additional heat to your laundry room. In addition to the extra heat, you will end up with a lot of unwanted moisture. Remember, bathroom, kitchen, and dryer vents must vent to the exterior (Fig. 3-16).

Chimney Defects

Chimney defects can allow moisture from flue gases and from external forces like rain or snow, to enter the building and add to condensation problems that already might

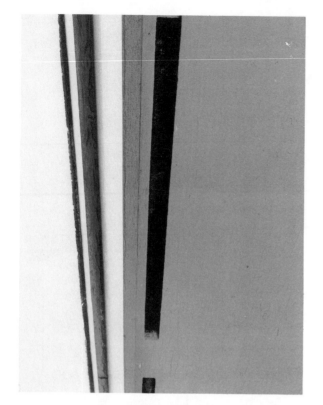

Fig. 3-5. Be sure soffit vents are not covered by attic insulation and that the screens are not sealed with paint. Either condition will result in poor or no ventilation.

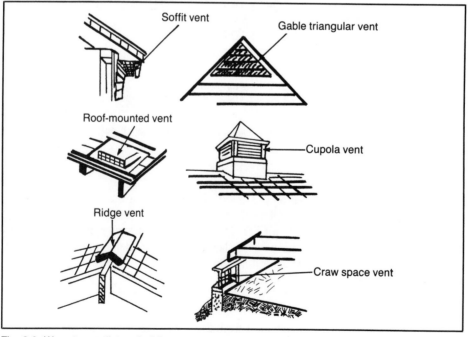

Fig. 3-6. Ways to ventilate a building.

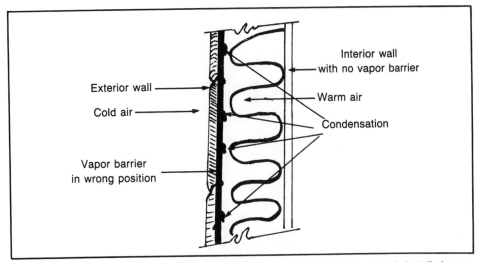

Fig. 3-7. Vapor barriers in exterior walls are crucial to a building's health. Incorrectly installed vapor barriers will do more harm than good, producing condensation that will damage both insulation and framing.

Fig. 3-8. Insulation in attics should have their vapor barriers facing down towards the areas below, and not facing up as seen in this photograph. The chance of trapped moisture and damage to the vapor barrier material is highly likely.

have a head start. Table 3-3 lists some of the common chimney defects that could directly or indirectly affect moisture buildup in your building. Be sure to address each one and, if any exists in your building, be sure to have proper repairs made. Figure 3-17 shows the correct height for chimneys, if yours doesn't measure up, have a mason make the necessary corrections.

Tight Buildings

Energy-efficient, "tight" buildings don't leave much room for trapped moisture to escape. This is particularly true of super insulated buildings that are well caulked and

Fig. 3-9. Crawl spaces with exposed soil need vapor barriers on the soil to control moisture movement in the crawl space.

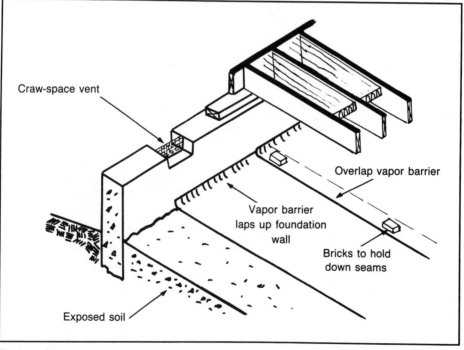

Craw-space vent

Overlap vapor barrier

Vapor barrier laps up foundation wall

Bricks to hold down seams

Exposed soil

Fig. 3-10. One way of controlling moisture movement in a crawl space. Lap all edges by several inches and provide some form of weight to keep seams down tight.

Fig. 3-11. Vinyl and metal sidings tend to hold moisture between them and the building framing. The end result usually is major decay and deterioration to the building. Additional ventilation is required for such buildings.

Fig. 3-12. Be sure to vent bathroom ceiling vents to the exterior.

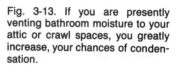

Fig. 3-13. If you are presently venting bathroom moisture to your attic or crawl spaces, you greatly increase, your chances of condensation.

Fig. 3-14. Never vent dryer vents into enclosed spaces such as attics or crawl spaces. The resulting moisture buildup will cause decay and damage to both framing and insulation.

Fig. 3-15. Avoid using heat reclaimers as seen on this dryer vent. The results in your laundry areas will be the same as those in enclosed spaces—decay to wood, peeling paint, and damaged framing.

Fig. 3-16. Vent appliances and fixtures directly to the exterior of the building.

weatherstripped. If you happen to own a supertight building, you probably are suffering from some form of condensation. If you don't see the obvious inside signs of moisture problems, it is somewhat likely that you could have some amount of concealed condensation trapped in the walls of your building. Exterior visual results of condensation might look like what you see pictured in Fig. 3-18. Usually the best solution for a very tight building is to have an air-to-air heat exchanger installed. See Table 8-5 in Chapter 8 for a list of manufacturers of these devices.

**Table 3-3. Common Chimney Defects
Affecting Moisture Buildup in Buildings.**

- ✓ Improper chimney height over roof lines
- ✓ Overhanging tree branches
- ✓ Overgrowth of ivy on chimney top
- ✓ Deteriorated chimney cap
- ✓ Damaged flue liners
- ✓ Caved-in flue
- ✓ Blocked flue by birds nests
- ✓ Openings in chimney walls
- ✓ Plugged up chimney weather cap
- ✓ Boiler flue too far into chimney
- ✓ Flue connections not sealed up tight to chimney
- ✓ Open chimney cleanout in basement
- ✓ Chimney needing a cleaning
- ✓ Deteriorated chimney portions in attic or basement
- ✓ Open chimney flashing
- ✓ Missing chimney cap

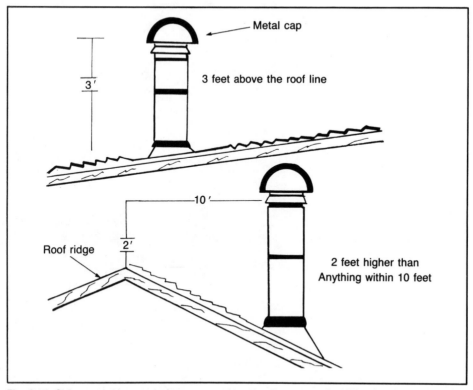

Fig. 3-17. Chimneys with poor drafts can sometimes be the cause of additional moisture problems in a building, particularly with gas-fired appliances. Be sure your chimneys are the correct height.

Fig. 3-18. Severe peeling of siding and trim areas as well as decay to wood are usually the results of buildings that are too tight and lack sufficient ventilation.

Overgrown Vegetation

Overgrown trees, shrubs, and vegetation that are too close to the building will tend to hold moisture and thus promote decay. Table 3-19 shows what can happen when you forget your landscaping maintenance chores. Be sure to trim back all overgrown vegetation to allow the building to breath and dry out from the suns rays. Failure to do so will not only induce condensation but will also promote decay and wood-boring insect activity.

Heating

Improperly vented heating devices and systems contribute moisture in the form of unvented combustion gases. Figure 3-20 demonstrates an improper open flue pipe which should be sealed up to keep flue gases from entering the building. Inefficient and dirty heating systems, as seen in Fig. 3-21, also contribute to improper combustion that, in turn, will result in excess moisture spilling into the building rather than venting to the outdoors. Be sure that you have your heating systems and space heaters inspected and serviced on an annual basis.

Humidifiers

If you want extra moisture in your building to aid condensation, then just buy yourself a humidifier. No matter if the humidifier is a free-standing one used for individual rooms or one that is installed directly into your heating system (Fig. 3-22), the results are the

Fig. 3-19. Allowing runaway vegetation to take over will block the sun's ability to dry out your building. Wettings from rain and snow will deteriorate various building materials. If you can't see your building because of overgrown vegetation, get out the old chainsaw.

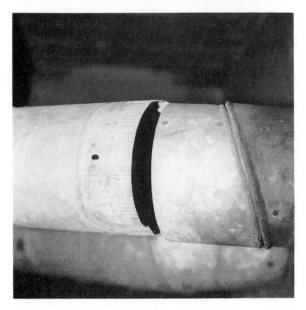

Fig. 3-20. Escaping combustive gases accumulating in your building are not only hazardous but bring unwanted moisture as well. Never allow such a situation to exist in your building.

Fig. 3-21. Dirty heating systems that need service means that the owner is not keeping up with maintenance chores. For a nominal fee you can have your heating system tuned up on an annual basis.

Fig. 3-22. A humidifier like this will cause more harm than good. The moisture that it pumps into your warm air system will, in time, rust out the heat exchanger as well as adjacent metal parts. If at all possible, avoid using a humidifier in a heating system. Consider a room unit instead.

same—extra, unwanted moisture being pumped into the building. An additional negative feature can be found in the furnace-mounted humidifier. In addition to pumping unwanted moisture throughout the building, it will, in time, also destroy your heat exchanger in the furnace. Moisture from the humidifier could rust out a heat exchanger in less than 7 years. Don't hesitate to remove it from your furnace, if you have such a hookup.

Windows

Windows that are loose, have deteriorated framing's or caulking, do not have storm windows, and are generally defective will also cause some of the factors involved in the condensation cycle. During your maintenance inspection tours make sure that all of your buildings windows are tight, weatherstripped, caulked, and puttied. If you have any windows without storm windows, (Fig. 3-23), be sure to add this updating to your "must do" list. If you do have storm windows but do not have weep holes, you could end up with moisture damage like that in Fig. 3-24. Note the discoloration from the excess moisture caused by lack of weep holes. If you have storm windows without these weep holes, take five minutes and drill two—one at each end of the sill-window joint.

Fig. 3-23. Single pane windows will result in both heat loss and moisture transfer from the exterior. If you presently have such windows, consider having storm windows installed.

Fig. 3-24. Storm windows without weep holes could cause condensation, resulting in discoloration and decay in trim areas.

Crawl Spaces

Poorly designed and maintained crawl spaces will, in a very short time, develop major symptoms of physical breakdown. Decay and wood-boring insect activity are usually the results of trapped moisture in crawl spaces. Damaged insulation as well as structural problems could also be a by-product of condensation in crawl spaces. Table 3-4 lists several recommendations that are useful in preventing the noted "ills" that your crawl space could have. An access panel (Fig. 3-25) will make it easier to get into your crawl space for maintenance checks and repair work.

Drying Things Indoor

Avoid drying firewood indoors. The stored-up moisture in the wood only contributes to the buildings existing moisture. Also, try not to dry clothes indoors on makeshift clothesline. Rather, use a vented clothes dryer or dry the clothes outdoors during the warmer months. After you take a shower, don't leave soaked towels hanging around; dispose of them directly into the hamper. The key here is to limit the amount of moisture that can get into the building's atmosphere.

Wet Basements and Crawl Spaces

As you should know by now, provide roof drainage that will discharge roof runoff several feet from the building. In addition, be sure to slope the foundation grade so that the water will flow away from the building rather than to it. Moisture in basements and crawl spaces from seepage and leaks will turn up eventually in the air as water vapor, and the vicious cycle of condensation will begin again.

CURES FOR CONDENSATION

As you have already seen in Table 3-2, there are many causes for condensation, and each one does have a remedy. Let's review for clarification some of the major cures that will help promote a drier and healthier building.

Table 3-4. Crawl-Space Recommendations.

Each of the following crawl spaces should have:

Heated crawl spaces
- ✓ Ground covers over exposed soil
- ✓ Insulation and vapor barriers on perimeter foundation walls
- ✓ Foundation vents closed during the winter

Unheated crawl spaces
- ✓ Ground covers over exposed soil
- ✓ Insulation and vapor barriers under floor areas
- ✓ Foundation vents open year round

Crawl spaces in warm climates
- ✓ Vapor barrier under flooring facing *down* towards the ground

Crawl spaces in cold climates
- ✓ Vapor barrier under flooring facing *up* toward the floor

Crawl spaces in all climates
- ✓ Ground covers of 6-mil polyethylene laid over exposed soil
- ✓ Seams and edges of ground cover held in place by heavy objects such as bricks or stones
- ✓ 4 to 6 inches of the perimeter edge of the ground cover turned up against the foundation
- ✓ Insulated heat and water pipes as well as insulated ducts

Ventilation

If there is one major cure for condensation, it would be ventilation. All buildings need ventilation, some more than others. A building with none or minimal ventilation is just begging for problems to develop. Figure 3-26 illustrates the many ways that a building can be ventilated. Try using some or all of them to help curb condensation in your building. Don't be afraid to open windows in the winter for five or ten minutes to air out the building. The heat loss will be minimal compared to the good that this will do, particularly if the building is showing signs of condensation.

Dehumidifiers

Use dehumidifiers in areas that show a high buildup of moisture. Basements are ideal places for them. A dehumidifier in a basement or crawl space will extract moisture from the air. You might be surprised to find that in as little as 24 hours a gallon of water can be collected from the air. Large buildings or buildings with major moisture problems will do best to have more than one dehumidifier.

Fig. 3-25. Crawl spaces need access areas to provide proper inspections and maintenance. If you have none, open one up so that you can do the required monitoring and upkeep.

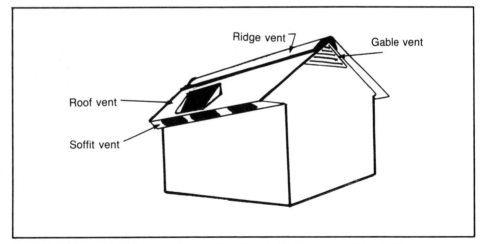

Fig. 3-26. Ways that buildings can be ventilated to get rid of excess moisture buildup.

Vapor barriers installed correctly and fitted tightly to wall, ceiling, and floor areas will curb moisture transfer from one part of the building to another part. If you have none, or if yours are damaged, be sure to either add or repair. If you find areas such as top floor ceiling that have insulation above them but no vapor barriers, try painting the ceilings with a vapor-barrier paint, which is quite effective. Vinyl wallpaper is also another good vapor barrier that can be effectively used to curb moisture movement through walls.

Ground Covers

Exposed soil in crawl spaces need ground covers that are actually vapor barriers. Sheets of 6-mil polyethylene installed correctly will prevent the transfer of moisture from the exposed soil into the crawl space. Make sure that all of the joints, including the perimeter, are sealed tight to prevent any moisture from escaping.

Vents

Bathrooms, kitchens, and laundry rooms all should be vented to the outside to avoid any buildup of moisture. Figure 3-27 shows mildew caused by excessive moisture in a bathroom. Make sure that your venting system does not pump moisture into attics, basements, or crawl spaces.

Chemical Desiccants

Desiccants, such as silica gel, activated alumina, and calcium chloride, are useful in helping to dehumidify basements, crawl spaces, and closets. These chemicals absorb the moisture in the air and usually work best in small confined areas, but they will work in larger rooms as well. Table 3-5 gives specifications for their general use. As with any chemicals, be sure to carefully read the directions on the packages.

Air-To-Air Heat Exchangers

To ensure a very tight, energy-efficient building you might want to install an air-to-air heat exchanger (Fig. 3-28). These mechanical devices ventilate and dehumidify tightly

Fig. 3-27. The results of trapped moisture in a bathroom. This mildew growth will, if left unattended, cause serious damage to the ceilings and walls in this bathroom.

Table 3-5. Chemical Desiccants for Dehumidification.

Silica Gel or Activated Alumina
- ✓ Use with clothes hanging in closets or place in containers on shelves in basements.
- ✓ Close off spaces to be dehumidified for maximum results.
- ✓ Not harmful to fabrics.
- ✓ Reuseable.
- ✓ Suitable for above-freezing applications.

Calcium Chloride
- ✓ Use in basements and crawl spaces.
- ✓ Close off areas to be dehumidified for maximum results.
- ✓ Corrosive to clothes and metals; causes burns to the skin.
- ✓ Not reuseable.
- ✓ Suitable for above-freezing applications.
- ✓ Keep away from vegetation.
- ✓ Wash surfaces after contact (e.g., metal, skin, clothes).

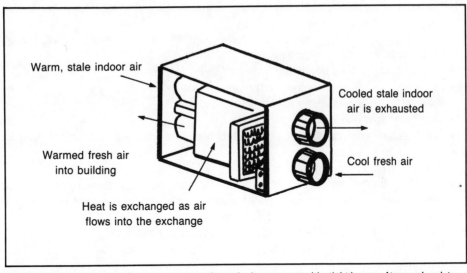

Warm, stale indoor air

Cooled stale indoor air is exhausted

Warmed fresh air into building

Cool fresh air

Heat is exchanged as air flows into the exchange

Fig. 3-28. Air-to-air heat exchangers are sometimes the last resort to rid a tight house of trapped moisture. Be sure that you really need one before buying one because most are expensive.

built structures, particularly those in cold regions. It works by recovering heat from outgoing stale air and uses it in turn to warm up fresh cold, incoming air. Before you buy one, however, contact a few reputable dealers to see whether you really need one.

Mechanical Tune-Ups

Have your heating and cooling systems as well as any space heaters annually inspected, cleaned, and serviced, Make any necessary repairs.

Insulate

Insulate heating and cooling ducts, heat pipes, hot and cold water pipes, hot-water tanks, crawl spaces, and attics. In insulating attics, it is important to do it right. Don't insulate the overhang where the soffit areas are. Insulation in this area could block the natural flow of air from the soffit vents or from cracks and open joints in the eaves. With the air flow blocked, moisture has a good chance of condensing, and you now know what that will mean.

SUMMARY

Indoor moisture in the form of condensation has many causes as well as many cures. During your ongoing maintenance program, be sure to pay particular attention for signs of condensation. Be especially alert in crucial areas such as attics, crawl spaces, basements, and exterior walls. Apply all of the suggestions given here in this chapter and review the checklist at the end of this chapter to help combat condensation. If you find that none of these measures are successful for you, then call in a professional for additional advice.

CHECKPOINTS

- Correct outside sources of moisture by draining off rain and melting snow away from your foundation walls.
- Reduce the source of excessive moisture in your building by repairing roof, siding, cellar walls, and faulty plumbing.
- Seal up leaks around doors and windows and also open joints.
- Install spot ventilation such as bathroom and kitchen vents.
- Exhaust appliances such as dryers to the outside.
- Don't dry firewood indoors.
- Cut back and trim up overgrown vegetation around your building.
- Avoid or limit the use of humidifiers.
- Tune-up your heating system annually.
- Provide plenty of ventilation in both crawl space and attic areas by installing a combination of ventilation units such as gable, soffit, ridge, and roof vents.
- Provide year-round ventilation in attic areas.
- Open up doors and windows to air out musty, damp basements.
- Provide ground covers over exposed soil in basements and crawl spaces.
- Use dehumidifiers if you have a damp, moist basement.
- Employ chemical dehumidification such as desiccants in closets and small rooms.
- Have an air-to-air heat exchanger installed in very tight buildings.
- Always be on the lookout for symptoms of condensation: sweating pipes, musty odors, peeling paint, frosted windows, unexplained dampness or moisture, and beads of water on window panes.
- Make sure vapor barriers (retarders) are in place.
- Monitor key areas such as the attic, crawl spaces, basements, and visible portions of exterior walls.
- If all else fails, consider professional help.

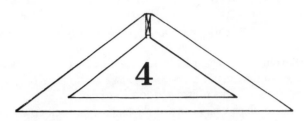

Roof Leaks

"Let a Smile Be Your Umbrella" was the title of a song some years ago, but it surely is no smiling matter when water comes through your roof causing decay to wood framing and damage to finished ceilings and walls. Water—as vital as it is to us—is a destructive force when it enters a building. As you have learned, water (rain, snow, ice) is the most serious adversary that you will face in your continuing maintenance program. In Chapters 2 and 3 you saw how water from the outside caused trouble in the basement and how condensation from within caused deterioration throughout your home or investment property. This chapter will help you cope with water that comes from the heavens to make your property a living hell. It will show you what maintenance chores you should do to make your roof weathertight.

Keep in mind that most roof leaks are not immediately obvious. As with the basement and the building throughout, you must pay heed constantly to prevent damage from roof leaks. You almost have to become a building detective. During the normal course of activities you must develop a keen sense of awareness as to changes taking place that are hardly noticeable. Is there peeling paint on walls or ceilings that were in good condition just recently? Are there cracks in plaster where there were none just a short time ago? Do you now smell musty odors or do some rooms now feel damp to you? Have you found stains on walls or ceilings or have you noticed peeling wallpaper? Many of these signs might be trying to tell you something. Don't be an oblivious owner who like the fabled three monkeys, albeit in a slightly different context: saw no moisture; heard no leaks; and felt no dampness.

Because you are an owner who is mindful of his/her surroundings and does not allow damage to go unchecked, read this chapter carefully. Remember, though, this book is not a repair book per se, but rather a preventive maintenance manual. In Appendix B you will find a comprehensive list of books on repairs that will not only help you with repairing roofs but with all sorts of other building repairs as well.

SAFETY TIPS FOR USING A LADDER

Because the inspection, maintenance, and repairs of roofs require some ladder work, it is a good idea for you to take some common-sense precautions. Here are some safety tips for working safely on a ladder:

- Never work in wet, windy, or cold weather. Most materials become dangerously slippery when wet.
- Make sure that your ladder is in tip-top condition. If it is seriously damaged, buy a new one.
- Always lean your ladder against a building wall so that the distance from the bottom of the ladder is one-fourth the distance between the base of the ladder to its top.
- Always rest your ladder on firm and level ground.
- If your ladder wobbles while you are climbing it, go back down and reset it firmly.
- Be sure to have someone hold the ladder while you are climbing.
- Keep metal ladders away from overhead electrical wires.
- Always face the ladder when going up and/or down and keep both hands on the rails, not on the rungs.
- Carry supplies or tools on your belt or hoist them up with a rope.
- When working on a ladder, never lean over in any direction. Keep your hips between the rails.
- Never climb over the top of a ladder to get to the roof. Get a longer ladder so you can step directly on it.
- If you fear heights—never climb ladders at any time!

FLAT ROOFS

The worst-designed roofs are those that are relatively flat such as a shed roof (Fig. 4-1) or a built-up roof (Fig. 4-2). Because these types of roofs do not shed rain or snow quickly, water ponds (Fig. 4-3) and/or ice dams can develop (Fig. 4-4).

Ponding water will eventually find its way into the building, the results of which are shown in Fig. 4-5. Ice dams, as you shall see later in the chapter, are not only caused by flat roofs but also by inadequate insulation and insufficient ventilation. Both of these causes are discussed in detail in Chapter 6. Suffice it to say that because they are such a problem in some parts of the country, a brief discussion as to how to cope or, better still, how to avoid them is included in this chapter.

To begin with, you might have trouble with a roof because it is flat, but there are other reasons for roof damage. For one, a roof deck might have been added at a later date. Very often these decks are poorly designed and also badly installed. A perfectly sound roof could begin to develop leaks after a new deck has been added on top of it. To avoid such a predicament, hire a contractor that will take all the necessary precautions

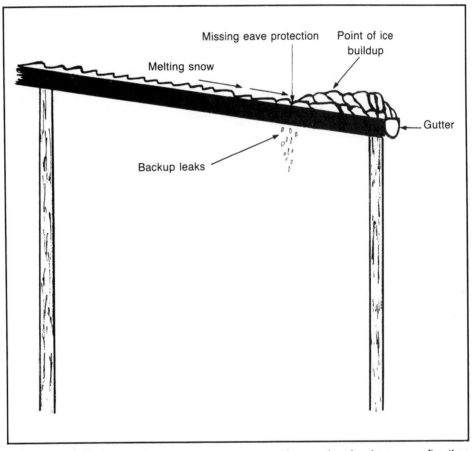

Fig. 4-1. Shed roofs, like the one here, tend to develop problems such as ice dams more often than steeply pitched roofs.

to preserve your roof. (See Chapter 10 to learn how to select a reliable contractor.) Table 4-1 lists the protective steps to take in adding a roof deck. Discuss these steps with your contractor.

Another way to damage a flat roof is to walk on it unnecessarily to enjoy leisure activities, such as sunbathing for instance, or perhaps for gardening. Figure 4-6 shows what happens when too many feet have been walking on roof coverings. The best solution is not to allow anyone to use a flat roof area for any purpose. Let the roof do only what it is meant to do, namely protect your building.

Table 4-2 lists the most common flashing materials available for roofs and roof components, such as chimneys and skylights. Experience has shown that one of the best types for the price and the ease of handling is lead flashing. You can really enjoy working with it because it can easily be molded to fit around intricate bends and turns. It can also be gently tapped so that it will take the form of the object being flashed, which in turn will make a roof component more watertight than if you used some other forms of flashing materials. In Figs. 4-7 through 4-10, you can see the simple steps to take when working with this type of flashing.

Fig. 4-2. Flat roofs tend to discharge water at a slower rate than pitched roofs, and hence some leakage into the building is likely—particularly if the roof is old and deteriorated.

Fig. 4-3. Ponding of water on flat roofs not only results in leaks but also, in some instances, damages the roof structure.

Fig. 4-4. On poorly pitched roofs, such as shed roofs, ice dams and ice damage will result under the right conditions.

Fig. 4-5. The results of ponding water and ice dams can readily be seen in this photograph. Peeling paint, damaged plaster, and decaying wood framing are the end results.

Flashing must be applied to all joints, openings, intersections, and to any areas where different materials join, such as where brick siding meets wood trim. Table 4-3 lists the most common places to install it.

Clearly, even though all roof openings (skylights, chimneys, plumbing vents, ventilation units, etc.) have been flashed, there is no guarantee that water will not leak through these areas. Eventually even the best flashing materials will deteriorate or open

Table 4-1. Tips for Constructing Roof Decks.

✔ Always use pressure-treated wood for the entire deck.

✔ Never place ends of support framing directly on the roof.

✔ Use weather-resistive padding between the framing and the roof.

✔ Make sure that the roof under the deck is in good condition prior to installing the deck.

✔ Try to install the deck so that, in case of any emergency, it can be picked up or removed to make repairs to the roof.

✔ Don't drive nails from the roof deck framing into the roof.

✔ Provide sufficient hand railings and braces for support and to prevent accidents.

✔ Don't cover important plumbing vent pipes or roof ventilation units with the deck.

Fig. 4-6. Roofs are not meant to be walked on except in emergencies such as repair work. Damaged finished surfaces are usually the result of unwarranted walking on roof surfaces.

Table 4-2. Flashing Materials.

Material	Cost	Application
Lead	Expensive	Very easy to work with
Aluminum	Inexpensive	Easy to work with
Sheet metal	Inexpensive	More difficult than lead
Vinyl	Moderate	Can't be used on all jobs
Copper	Very expensive	Used on top quality jobs

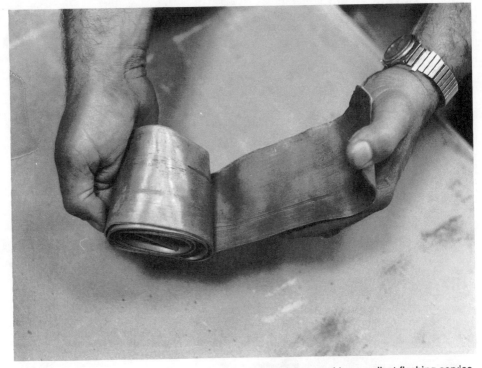

Fig. 4-7. Lead flashing is a relatively heavy-gauge material that provides excellent flashing service.

up to allow water in. Figure 4-11 shows just such an example. To protect your building from roof leaks, you have to inspect it on a regular basis—in the spring and fall. Check those areas listed in Table 4-3 and provide regular maintenance in the way of sealing up open gaps with roofing cement.

If maintenance does not do the trick any longer—for example, the flashing is damaged beyond repair—the flashing will have to be replaced. This is usually a job for a professional; however, if you do want to tackle such a replacement, read this excellent manual entitled, *Residential Asphalt Roofing Manual* (see Appendix B).

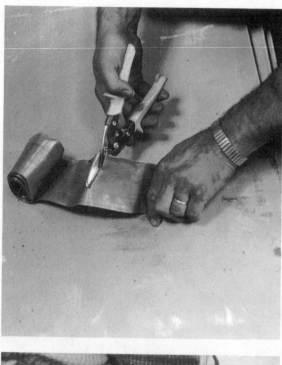

Fig. 4-8. Lead flashing can be cut to size and shape with metal snips.

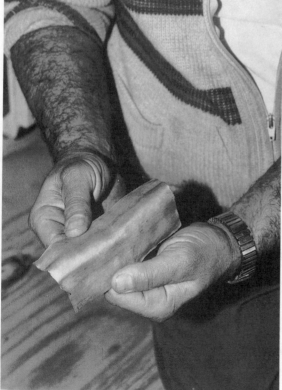

Fig. 4-9. Because this material is so pliable, it can be bent and shaped to meet almost any configuration.

Fig. 4-10. Lead flashing gives a tight, waterproof, sealed joint that will give years of trouble-free service.

Fig. 4-11. Inspect and maintain flashing areas to avoid open areas such as this, which will result in major leaks to the building.

Table 4-3. Where Flashing Should Be Applied.

- ✔ Chimney/roof lines
- ✔ Vent pipe roof line
- ✔ Flat roof perimeters (edges)
- ✔ Roof valleys
- ✔ Dormer valley and ridge lines
- ✔ Intersecting roof and wall lines
- ✔ Between dissimilar materials
- ✔ Over window openings
- ✔ Over door openings
- ✔ Around ventilation units
- ✔ Around skylight openings
- ✔ Edges of roof areas such as eaves and rakes (drip edges)
- ✔ Any opening in a building through which water can penetrate, such as the previously listed areas

Table 4-4. Roofing Materials.

Type	Relative Cost	Durability
Asphalt/fiberglass shingle	Inexpensive	15-25 years
Roll roofing	Inexpensive	10-15 years
Built-up roofing	Moderately expensive	10-20 years
One-ply membranes	Moderately expensive	10-20 years
Wood shingles	Moderate to expensive	15-30 years
Wood shakes	Moderate to expensive	25-50 years
Metal	Moderately expensive	25-50 years
Clay tiles	Expensive	50-100 years
Slate	Expensive	50-100 years

ROOFING MATERIALS

Table 4-4 lists some of the most common roofing materials that are used across the country. Relative cost ratings and typical life expectancies are given for comparisons. When reroofing, also ask your roofer for his opinion on what to use for replacement roofing materials.

No matter what type of roof covering you have, the first step of your preventive maintenance program is to visually check for potential trouble spots and/or active leaks. Check Table 4-5 for likely areas for leaks. Make sure to go through the inspection ritual

Table 4-5. Likely Areas for Roof Leaks.

✓ Flashing areas around chimneys, vent pipes, skylights, etc.
✓ Deteriorated or damaged shingles or roof coverings
✓ Damaged chimneys
✓ Vent pipes
✓ Roof drainage systems
✓ Skylights
✓ Valleys
✓ Hips and ridges
✓ Dormers
✓ Ventilation units
✓ Damaged siding or trim
✓ Ice dams
✓ Exposed nails
✓ Intersecting roof lines
✓ Open seams in roofing materials
✓ Missing drip edges or flashing

on a regular basis. If water is already leaking through your roof, find the source of entry as soon as possible to avoid further damage. In your search for leaks remember the following factors, all of which are working towards the demise of any and all roofing materials:

- A building settles at a different ratio than masonry chimneys. This differential settlement causes flashing to pull away from chimney and roof connections.
- Given time and exposure to weather, even the best caulking compound and roof cement will dry out and leave open joints.
- Some forms of metal flashing will in time corrode and rust through.
- Exposure to sunlight will cause asphalt or tar to blister, crack, and deteriorate.
- Sealed joints will eventually pull apart because of natural expansion and contraction of materials caused by heat and cold weather.
- Untreated wood will eventually decay and deteriorate.
- Mold and mildew will in time break down roofing materials.
- Wind and flying objects, such as tree limbs, will cause damage to roofing materials and roof components.
- Antennas attached to chimneys and vent pipes are prime sources for roof leaks.
- Improperly installed roof decks will cause damage to the roof which, in turn, can become a source for leaks.
- Careless walking on roof coverings can damage them.

HOW TO FIND ROOF LEAKS

Finding roof leaks is far from easy and sometimes requires the skill of a detective. Try to trace the leak first from interior signs; remember that even a slight discoloration in a wall or ceiling suggests moisture. Remember, though, that the actual point of entry might be miles from where you find a stain. Melting snow could enter through a defective shingle, for example, run down a rafter, drip onto the attic floor, and flow several feet before it drips down into the living room below. See Fig. 4-12 for such a flow pattern. Such leaks are difficult to find, at best. You might not be able to detect it unless you wait until it is raining out, then the search begins.

During a rainstorm or during the time when temperatures are mild enough to melt snow on your roof, take a flashlight and seek out an honest leak. Station yourself in the general area of discoloration in the wall or ceiling, turn the light off and listen for dripping water. Run your hand carefully along the roof areas that appear to be affected and see if you can feel moisture. Once you detect the spot where you think the water might

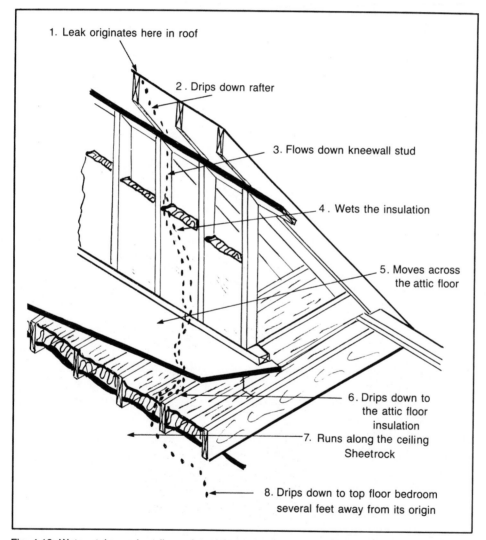

Fig. 4-12. Water stains and peeling paint might be the result of a leak that originated several feet from the visible results. Check attic and crawl spaces during heavy rains to determine the origin of the leak.

be coming in, mark it with a pen or marker. Later, when the weather has cleared up, double-check the adjacent exterior areas. Chances are that you probably will find the damaged or deteriorated part—a damaged shingle, and exposed nailhead, or deteriorated flashing, for instance.

SAFETY TIPS FOR WORKING ON ROOFS

Once you have located the source of the leak, the repair work could be quite simple. But before you rush up onto your roof, follow the roof safety tips given below:

- Never work on roofs if you have a fear of heights.
- Be sure to follow the Ladder Safety Tips when climbing ladders.
- Avoid working on roofs during inclement weather, particularly during rain, snow, heavy winds, or during lightning storms.
- Never work on roofs when they are wet from rain, snow, or heavy dew.
- If you are using a scaffold, follow the assembly and usage directions provided by the dealer.
- Check the soundness of scaffolds before using them.
- On steep roofs use a safety line anchored to the opposite side of the building.
- Always wear rubber-soled work shoes or sneakers.
- Avoid any contact with electrical wires or power lines.
- Do not place heavy loads of roofing materials on one spot. Distribute the roofing materials over the roof.
- Check to make sure that the roof will support the weight of new roofing materials. If in doubt, get professional help.
- Get help in lifting heavy objects.
- Always take your time in dangerous situations; walk carefully.
- Keep your work area clean so you won't trip over anything.

Besides using the safety tips, be sure to follow the list of *do's and don'ts for working on roofs*:

- Don't slam ladders against gutters as you could easily damage them.
- Only apply asphalt shingles or roofing materials on a day when temperatures will allow laying them out flat.
- If repairs are needed during cold weather, keep shingles in a warm, dry location until you are ready to use them.
- Always use a sealant that a manufacturer recommends with their products.
- Always cover exposed nailheads with asphalt cement.
- Always use the correct size and type nails that the manufacturer of the roofing material recommends.
- If you have to make emergency repairs during cold weather, be sure to keep roofing cement in a heated area until you are ready to use it.
- When replacing gutters, always give them a slight pitch towards the downspouts. A 1-inch pitch for every 16 feet is an accepted industry standard.
- Always store roll roofing in an upright position in a dry, warm area, but never near heating equipment.

- Store cans of roofing cement away from heating systems.
- Protect roofing materials from freezing weather. Moisture freezes in cold weather and can damage roofing materials.
- Always replace the lids on cans of asphalt cement when you are not using them.
- So you won't crack or damage roofing materials, never walk on a roof when the temperatures are very cold or very hot.
- Always follow the safety rules suggested in this book in addition to your own common-sense precautions.

SIMPLE ROOF REPAIRS

Temporary roof repairs can be made by any handy person. Probably one of the simplest repairs is to apply a good dab of roofing cement to openings in roofing materials or to defective flashing areas. It is actually a good idea to always have a full 1-gallon can of roofing cement on hand for quick repairs. A relatively new twist to roofing repairs with asphalt cement is the use of a cartridge of roof cement and a caulking gun. Instead of digging roof cement out of a can with a trowel, all you have to do is to squeeze it out from the caulking gun (Fig. 4-13).

There are many different types of good sealants and patching compounds available to the public. Each of the four kinds listed in Table 4-6 has different attributes and

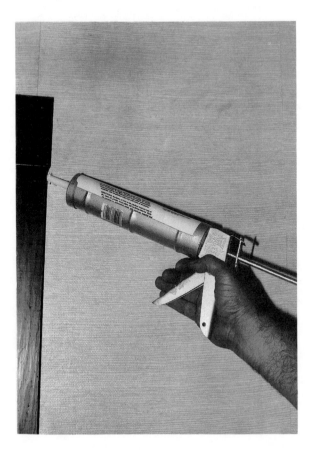

Fig. 4-13. Use a caulking gun to apply caulking compounds easily and efficiently.

Table 4-6. Types of Sealants and Patching Compounds.

Asphalt Cement
- ✓ *Color*: black
- ✓ *Application*: by caulking gun or by trowel
- ✓ *Repairs*: seals cracks in underlayment, seals cracks in roofing, glues down shingles, seals flashing, patches damaged shingles
- ✓ *Advantages*: can be bonded to asphalt, metal, or wood

Butyl Caulking Compound
- ✓ *Color*: white or off-white
- ✓ *Application*: by caulking gun or putty knife
- ✓ *Repairs*: used to seal narrow cracks, patches cracks and holes in shingles, seals joints in flashing, seals cracks in mortar, seals the edges of flashings
- ✓ *Advantages*: can be painted over; stays flexible long after application

Aluminized Sealer or Caulking
- ✓ *Color*: aluminum
- ✓ *Application*: by caulking gun or trowel
- ✓ *Repairs*: used to patch and seal sheet-metal materials, such as gutters, downspouts, flashings, and skylight housings
- ✓ *Advantages*: can be used during rain because it bonds to wet surfaces

Clear Butyl Sealer
- ✓ *Color*: clear
- ✓ *Application*: caulking gun
- ✓ *Repairs*: makes waterproof joints on plastic skylight domes
- ✓ *Advantages*: Can be bonded to wood, glass, concrete, and metal

advantages. All of these materials can be found in any good hardware or supply store. Keep in mind, though, no matter which type of sealant you use, be very liberal in your application to the areas that you suspect are damaged and those that are clearly deteriorated.

For temporary repairs of larger openings, a piece of shingle or sheet metal should be cut to fit over the damaged area. The patch material is then slipped under the adjacent roofing materials and sealed in place with roofing cement or one of the other sealant listed in Table 4-6. This temporary patch usually can make a fairly long-standing seal until such time as a full repair can be made. Just make sure that you record on your master maintenance schedule that this temporary patch will have to be attended to in the near future.

CHIMNEY LEAKS

When you think of roof leaks, somehow it is hard to imagine that a chimney could be the culprit, but in many instances it is. Leaks can occur from open deteriorated flashing, damaged mortar joints, defective chimney caps, deteriorated or missing bricks, and/or rusting metal chimneys.

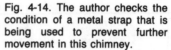

Fig. 4-14. The author checks the condition of a metal strap that is being used to prevent further movement in this chimney.

Figure 4-14 shows an example of how a chimney has moved away from a building, resulting in an open seam. A seam often plays the role of a highway for water and wind to travel on. A shifting away from the building should be monitored very carefully, and if the seam appears to get larger as time goes on, it should be corrected by attaching steel straps around it. These steel straps are bolted into the frame of the building to prevent further movement of the chimney. The open gaps can be caulked and sealed with any good quality caulking compound, such as butyl. The reason as to why a chimney takes a journey of its own away from a building is most likely to be found in a structural defect, such as a poor footing.

Perhaps the worst thing that you can do to your chimney is to attach an antenna to it. Unseen by anyone, the wind makes the antenna act like a sail with the chimney playing the part of a boat that is being moved about. Naturally, the flashing seal won't last; bricks will crack and deteriorate as will the mortar in the joints. Water will quickly find its way through such openings. So, if you have an antenna mounted on a chimney, do yourself a favor and remove it at your earliest convenience. The pleasure of a clear TV image is not worth the nasty image of water leaking into your building and causing all sorts of damage.

Table 4-7 gives you some simple pointing up tips for making repairs to deteriorated brick joints. To make sure that your chimney is waterproof, follow these helpful hints:

- Remove antennas from chimneys.
- Caulk and seal open joints in flashing areas.
- Point up brick joints.

Table 4-7. Pointing Up Tips for Chimneys.

✓ Safety first. Wear soft-soled shoes, use safety brackets with ladders. If heights frighten you, have someone else do the job.

✓ Rake (scrape) all loose mortar and debris out to a depth of at least 1 inch.

✓ Follow up the raking with brushing out of loose particles from affected joints.

✓ Hose down joints to be pointed up.

✓ If no hose is available, brush down with a wet brush.

✓ Follow the manufacturer's directions for mixing the mortar mix.

✓ Only mix up enough mortar for the job.

✓ When applying, don't be afraid to really pack the mortar into the open joints.

✓ Use a convex striking tool to strike the mortar joints.

✓ Strike the vertical joints first, then finish off by striking the horizontal joints.

✓ Press your tool into the joints so that you end up with a concave joint.

✓ Hose down the struck joints with a light mist to give them added strength.

✓ For three consecutive days, apply a light mist to the struck joints.

✓ Wash off any mortar spills with muriatic acid.

✓ Be sure to read the directions and follow them carefully when using any form of acid.

- Repair cap areas.
- Replace missing or damaged bricks.
- Replace damaged or rusted metal chimneys.
- Caulk and seal open joints between an exterior chimney and the building siding.
- Be sure to have your chimney flue liners cleaned and inspected regularly by a professional chimney sweep.

VENT PIPES

Other roof openings at which leaks are likely to occur are those for vent pipes. Here the cause for leaks is usually defective flashing at the base of the vent pipe or movement of the vent pipe by an antenna that is attached to it. Both problems are simple to solve: use roofing cement to seal up the deterioration at the flashing and remove the antenna from the vent pipe. Once the repairs are made, be sure to monitor the area under the vent pipe to see if your repairs are holding up. The best time to check is during a rain storm. Figure 4-15 shows an antenna attached to a vent pipe. Very poor judgement, indeed!

SKYLIGHTS

It does not really matter whether your skylights are old or new, you can usually depend on some trouble with water around them. Either moisture from inside vapors forming condensation, or melting snow and rain sneaking by the flashing, or a combination of both could cause skylight problems. The best solution to condensation, as was stated

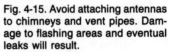
Fig. 4-15. Avoid attaching antennas to chimneys and vent pipes. Damage to flashing areas and eventual leaks will result.

in Chapter 3, is to provide adequate ventilation in the building. If you are dealing with a leak from the outside, first check the upper portions of the skylight that face the slope of the roof because this area is hit hard by rain and melting snow. Figure 4-16 gives you a good idea why this skylight is leaking. If you have a skylight that looks like this, tap the flashing back down and seal it up nice and snug with roofing cement.

Older skylights can succumb to water because of defective glass, deteriorated putty, or rotting wood. Use common sense here—if you think it is worthwhile to repair, do so. If the skylight is too old, however, or the damage too severe, simply replace it. This usually solves not only the leaks, but also provides you with a much better thermal barrier in the form of double-pane glass.

Helpful hint: Plastic or Plexiglass skylight domes that are cracked can be sealed with a clear butyl caulk that is meant to be used expressly with clear plastics. While patching, try to mend the crack from the inside as well as the outside. The outside patch should overlap the crack at least ½ inch on each side.

VENTILATION UNITS

Roof leaks can occur from almost any roof opening, and certainly roof or attic vents are no exceptions. There are other vents such as ridge vents, gable vents, and soffit vents, all of which are illustrated in Chapter 6 where ventilation is discussed. For now, all you need to know is that attic or roof vents are set into the roof or upper walls of the building, a fact that makes it quite easy for water to enter through defects in those

Fig. 4-16. The flashing is not under the shingles as it should be. The results could be leaks to the building.

ventilation units. So be sure to inspect these areas very carefully. Do it inside during periods of high winds and driven snow or rain. Look for signs of water penetrations. If you see anything in the way of moisture, wait until the weather is good and then check the flashing around the ventilation units from the outside. Make all the necessary repairs, and if you find a unit that is damaged, repair or replace it before additional wettings occur.

Vent tips:

- Check the condition of screens to make sure that no flying insects can come calling.
- Repair louvers to be sure that no driven snow or rain can enter the attic areas.
- Never walk on the ridge if you have a ridge vent.

ROOF DRAINAGE

Unless your home or building has 2-foot overhangs and sloping grounds that flow away from your foundation, you will definitely need a roof drainage system in the form of gutters and downspouts. They are necessary even though they are aggravating at times—aggravating in that they do need maintenance and service, usually more than once a year. Gutters clog up easily with leaves and debris while downspouts are sometimes inadvertently damaged by a lawnmower gone astray or by playful feet. Maintenance and upkeep of a roof drainage system are of supreme importance, though, not only preventing roof leaks but also, as you have already seen in Chapter 2, in preventing wet basements.

If you live in a region where the chances are high that leaves and twigs are blowing around which in turn end up in your gutters, you must get rid of this debris at least twice a year—in the spring to clear away the winter flotsam and in the fall to get ready for the onslaught of winter. Sometimes it might be necessary to clean gutters more often to get rid of the last of the fall leaves.

One handy way to prevent your downspouts from clogging up with leaves and other debris is by installing a screen plug. If the screen plug clogs up—which it periodically does—it is an easy and quick chore to clear it up. This type of screen is especially important on flat roofs that have center drains. Figure 4-17 shows a typical flat roof drain without this safety device. Not only will leaves drain into this opening but also roof gravel, balls, and whatever else you can imagine, sometimes with disastrous results.

Sometimes building owners who have had such problems resort to installing a "gooseneck" on the roof vent pipe, as seen in Fig. 4-18. Goosenecks not only obstruct the path into the pipe, but they also prevent vandalism to the system. It is not unheard of that juveniles decide to get rid of their empty beer cans by just dropping them down the vent pipe. The resulting repairs to the pipe and adjacent walls, as you can imagine, are quite expensive.

Quite often plugged up gutters and downspouts will back up during a heavy rain with the result that you now have to cope with a roof leak. If you have gone through such a disaster, you should consider for future repairs the addition of a protective 3-foot eave barrier. Figure 4-19 illustrates this simple roofing trick, which is also helpful in preventing damage from ice dams.

Maintaining gutters is as important as cleaning them regularly. Wood gutters tend to decay at joints and at the ends where they and the downspouts join. Left unattended

Fig. 4-17. Flat roofs without screen covers for drains will, in time, accumulate more than water in the drains.

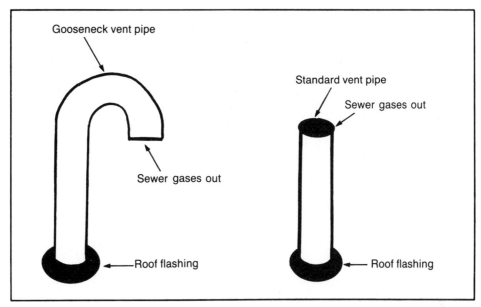

Fig. 4-18. A gooseneck vent pipe will prevent unwanted items from entering the vent pipe system.

Fig. 4-19. A metal roof edge is very helpful in preventing ice dams from forming. The 2- to 3-foot width prevents snow and ice from clinging to the roof.

it will spread to adjacent wood areas with the end result being major replacements. A simple maintenance chore to avoid such rot is to periodically (spring and fall) oil the inside of the gutters. After the gutters are cleared of accumulated leaves and other debris, liberally soak the inside wood with a 50%-50% solution of linseed oil and turpentine. Use either a rag or brush to work the mixture into the wood. Let the gutter soak up the mixture, then a few hours later, repeat the process. This should be done every 6 months.

To stretch out the life spans of damaged joints, apply roofing cement to the damaged areas, then cut a piece of metal flashing (lead flashing works very well) and press it into the cement. Once the flashing is in place, spread more roof cement over the exposed portions of flashing to make a nice tight seal.

Metal or vinyl gutters need less attention than wood gutters. Every 6 months make sure that the hangers that secure them to the building are fastened tightly to the building. Check the pitch of all gutters to be certain that they drain in the direction of the downspouts. If you find that during heavy rains, leaks occur at the joints, make a note for future caulking and sealing with a good quality gutter caulking compound.

ICE DAMS

Figure 4-20 shows a typical ice dam problem that affects many homes in cold climates. If you ever had to put up with such a problem, you will want to know how to avoid it.

To eliminate the possibility of an ice dam forming on your roof, you will have to do three things:

1. Insulate the attic areas thoroughly.
2. Ventilate the area sufficiently.
3. Provide proper protection along the eaves.

In Fig. 4-20 you can also see how ice dams can be avoided. Snow that has accumulated on your roof during a snowstorm is melted by the heat that is escaping through an uninsulated attic floor. The melting snow runs down the roof until it reaches the areas over the soffit. Because the soffit is colder than the roof areas directly over the house, the water freezes and forms a dam. More water accumulates behind the dam, backs up, and seeps down into the attic or rooms below. This dripping water will ruin ceilings, cause irreparable damage to insulation, and help to rot framing members.

By preventing the heat loss through the attic floor, you have to insulate these areas completely. This should solve 90 percent of the ice dam problem. Attic floors should have at least 6 inches of insulation, more in very cold climates. Attic insulation will probably be one of your best investments. It will also help save energy dollars, as well as prevent ice dams.

Because insulation does not completely solve the ice dam problem, you must provide sufficient ventilation for the attic. Usually a combination of soffit and ridge vents (Fig. 4-21) will help keep the air in the attic dry and at a temperature slightly above that of the outdoors. A well-ventilated attic reduces the chances of snow melting on your roof. See Chapter 6 for a full explanation of ventilation and how to use it to the best advantage of your building.

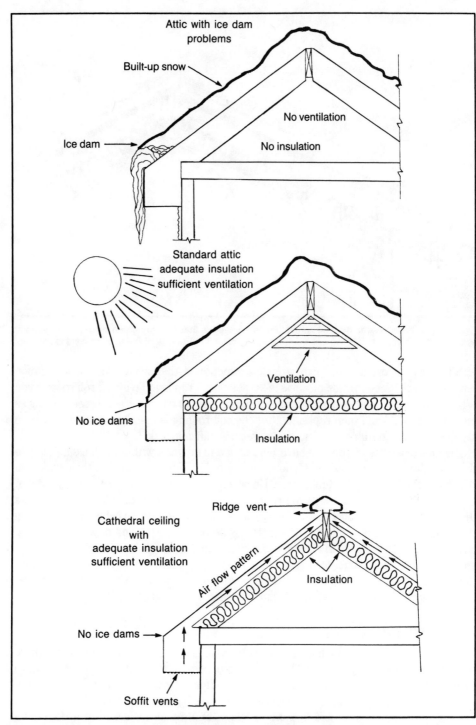

Fig. 4-20. How ice dams form. Proper insulation and ventilation will prevent such a calamity.

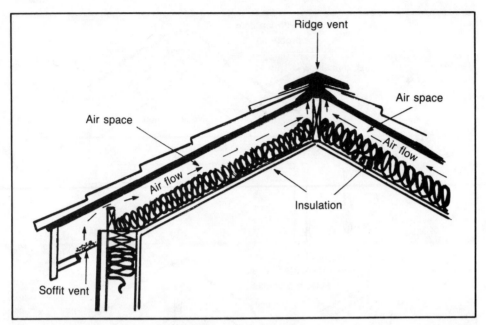

Fig. 4-21. Probably one of the best ventilation combinations that a building can have is a ridge-soffit system. Remember that this system won't be as effective if either one of the two partners is missing.

The last important step is to provide protection at the eaves. The most common type of material used for this purpose is 90-pound roll roofing. To install roll roofing you will have to remove the existing shingles at the eaves. Usually several courses of shingles must be taken up and then replaced after the roll roofing is installed. You should have a roofer do this job for you as it is a somewhat difficult task.

If for some reason you still have chronic ice dam woes with your building despite the insulation, the increased ventilation, and the roll roofing protection at the eaves, you might want to try using heating cables. Usually it is best to use two sets for each eave. One set of heating cables keeps the roof overhang warm while the other set prevents the gutters and downspouts from freezing up. While the first set of cables zigzags along the roof line, the other set runs along the inside of the gutter and down through the downspout. Both are held in place by metal clips that come with the cable. In Fig. 4-22 you can see "Harry Homeowner's cables." The problem here is two-fold. For one, the cables do not cover sufficient surface areas because they are too spread out; and second, there are no corresponding cables in the roof drainage system.

INTERSECTING JOINTS

Wherever different roof lines meet, with a wall or with each other (Fig. 4-23), there is always the possibility of a roof leak. This is especially true if the flashing has been neglected or is old, missing, or damaged. Sometimes, by mistake, the flashing is not installed at a roof/wall connection, and the results are highly predictable. If the affected section can be opened up and flashing installed, then by all means have it done by a professional. If the area is difficult to work on, however, then apply a liberal coat of roofing cement to these affected areas on an annual basis.

94

Fig. 4-22. For heating cables to be effective, space them close enough to melt snow and ice and use them in conjunction with heat cables in the roof drainage system (gutters and downspouts).

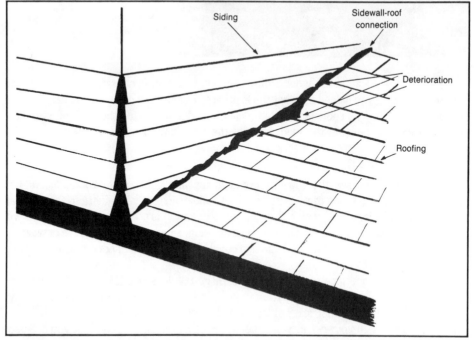

Fig. 4-23. Intersecting joints in a building, such as between walls and roof areas, need sufficient and properly installed flashing to prevent leaks to the building.

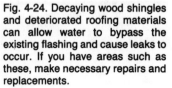

Fig. 4-24. Decaying wood shingles and deteriorated roofing materials can allow water to bypass the existing flashing and cause leaks to occur. If you have areas such as these, make necessary repairs and replacements.

Usually it is decay in the wood siding or wood trim that is causing the problem and not the flashing. If you indeed find that the leaks are coming from decayed and deteriorated wood, replace them as soon as possible. Figure 4-24 demonstrates the type of extensive deterioration that became the pied piper of the marching army of droplets.

HIPS AND RIDGES

Hips and ridges are the top cap areas of hip and gable style roofs. Figure 4-25 illustrates these two types of roof styles. Roof leaks frequently originate in these areas for a number of reasons of which the two most common are exposed nails (which will be discussed) and damaged cap shingles. Hips and ridges get damaged because of weather conditions and exposure to the sun. Damage could also occur by someone (roofer) walking on them.

In any case, once hips and ridges are damaged it does not take long for water to find its way through them. During your inspection tour, scrutinize these areas closely, and if you find shingles that are not up to par, replace them.

To help you with your repair work, here is another excellent book that you can use. It is called *Roofing Simplified*, (see Appendix B) and explains specific roofing repairs.

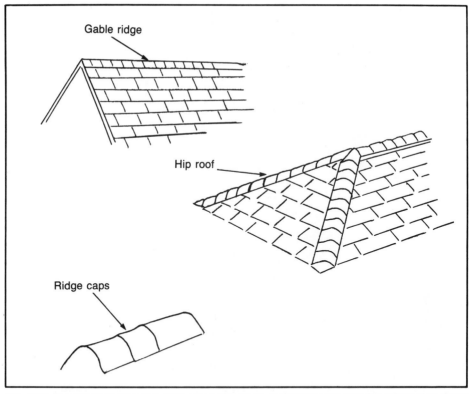

Gable ridge

Hip roof

Ridge caps

Fig. 4-25. Cap areas of hips and ridge areas should be periodically checked for defects that might allow water to penetrate the building. Exposed nailheads can be sealed up with a dab of roofing cement.

VALLEYS

A valley, or part of the roof at which joints of roof framing members converge, is another likely area where leaks originate. In Fig. 4-26 you can see the valley rafter in an attic; notice the water stains from active leaks in the valley areas. Valley repairs should be left to a professional roofer. Temporary repairs, however, can be done by a good handyman. Use asphalt roofing cement to seal up the exposed damaged areas in the valley. Remember, this is only a temporary patch that might be good for a short period of time only. Mark it down on your maintenance schedule for future professional repairs.

EXPOSED NAILHEADS

The roof surface is exposed to a variety of potentially harmful weather conditions. Over a period of time, the heat of the sun causes asphalt shingles to blister, curl, cup and dry out, and become very brittle. Shingles split or crack because of temperature changes, and strong winds blow away portions of deteriorated shingles. The exposed nailheads then become a potential avenue for water penetrations. A quick, temporary repair is to lay a dab of roof cement over the exposed nailhead and later replace the damaged shingle(s). If you have a roof that looks anything like that seen in Fig. 4-27, run to your phone and call a roofer.

Fig. 4-26. During and after heavy rains, check the attic valley framing areas for signs of water. In this photograph you can see where roofing cement has been applied to keep out water.

Fig. 4-27. Advanced stage of roof failure. This roof is well beyond its useful life expectancy. Leaks to the interior have probably been occurring for years.

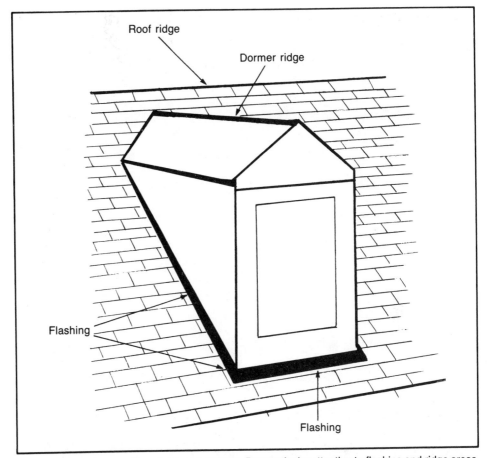

Fig. 4-28. Treat dormers like miniature buildings. Pay particular attention to flashing and ridge areas.

DORMERS

Dormers should be treated like any other projection on the roof. Like chimneys, vent pipes, skylights, and the like, dormers break the normal flow of the roof. A typical dormer can be seen in Fig. 4-28, notice that it is like a little house. Water can seep through the top from the ridge, from valley areas, through defective flashing, and intersecting roof lines. If during your inspection tours you notice stains on wall or ceilings in your dormers, check those listed exterior areas for signs of deterioration or damage. Be sure to make the necessary repairs.

SIDING AND TRIM

Quite often stains appear on ceilings and walls, and the blame is immediately put on the roof and/or roof components, such as chimney flashing areas. Many times the leaks have absolutely nothing to do with a defective roof. Missing or damaged siding and trim (Figs. 4-29 and 4-30) can often allow water penetrations to the building. So, once more check not only roof areas but also the entire complex structure of your building to find the source of the water penetrations.

Fig. 4-29. Mysterious leaks into the building might not be the result of a defective roof. The entrance area could be the siding.

Fig. 4-30. Damaged and decaying wood trim can sometimes be the entry point for water penetrations.

An excellent book for information on repairs to roofs and siding is simply entitled: *Roofs and Siding* and is listed with many other fine books in Appendix B. For specific directions on repairs and replacements for roofs as well as other building parts, build your own personal library so that you can have quick answers at your finger tips.

SUMMARY

As you must know by now, not all roof leaks are found in the roof per se. This chapter has attempted to show you where leaks are likely to occur and also how to make some simple maintenance and emergency repairs. Roof leaks can be pretty much avoided and damage curtailed by carefully inspecting your roof, taking careful notes for your maintenance schedule, and following through with simple repair work. Reread the pages and go over the information given in the tables. If you follow all of the steps outlined in this chapter, you should be smiling when it is pouring out because you know that the umbrella covering your building is watertight and free from leaks.

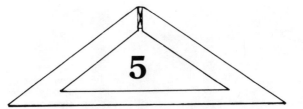

Decay and
Wood-boring Insects

Wood is probably one of the finest building materials that we have today. Countless older homes and estates that can be seen in most cities and towns are a standing testimony to the durability of wood. Figure 5-1 shows a house built in 1720, certainly a telling example of long-term good performance and durability. In Figure 5-2, however, you can see that not all wood-framed buildings fare so well, most likely because this home was poorly maintained over the years. In fact, the results of letting a house go, as it were, can be quite disastrous.

Over a period of time, wood deteriorates biologically, as does most everything that lives or has lived. It also succumbs easily to the work of wood-boring insects. In this chapter, you will learn more about the causes of wood decay and what you can do to delay the process. You will also learn about the types of wood-boring insects that thrive on wood and the necessary steps for you to take so they will not be able to destroy your property. If a building is kept free of decay and wood-boring insects, it should last several hundred years!

BIOLOGICAL DETERIORATION

The dictionary defines *decay* as a gradual deterioration from a sound condition to a general rotting or chemical decomposition. Stop and think about that! Is your home or investment property rotting away so gradually that you don't realize it? If it is, then you share a common problem with millions of other owners.

Biological deterioration is caused by both microscopic and visible plant life. The most common are fungi, bacteria, algae, mosses, and lichens.

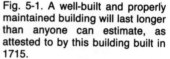

Fig. 5-1. A well-built and properly maintained building will last longer than anyone can estimate, as attested to by this building built in 1715.

Fungi are made up of four major groups: mold, stain fungi, decay fungi, and soft-rot fungi. These primitive plants do not have roots, stems, or leaves and cannot manufacture their own food like most other plants do. Rather, they draw nutrients from the cellulose materials found in wood. Figure 5-3 illustrates such a cycle that causes wood to decay: air borne fungi spores contact moist wood, draw nutrients from the wood, and, in time, produce new spore bearing structures called fruiting bodies.

Bacteria, like fungi, attack and damage wood in much the same manner, causing a progressive loss of strength and density in the wood fibers that in turn increases the likelihood of water being absorbed by the wood. Unlike fungi, however, bacteria become less active—particularly once the wood has become part of a building structure—and thus less of a concern for you.

Unlike fungi and bacteria, *algae, moss,* and *lichens* grow on the outside of the wood framing for support and as a supplier of moisture. The damage caused by these plants is confined to surface discoloration. Their presence in turn fosters the likelihood of fungi growth, however. By the way, a solution of sodium hydrochloride mixed with water will readily remove these plants from their nesting sites. A hardware or garden supply store should be able to supply you with this chemical. As always, read the manufacturer's directions carefully before application.

Fig. 5-2. A poorly maintained building will deteriorate long before it should. Remember, maintenance is the key to a healthy building.

Probably the most commonly known fungus is *mildew*, which feeds and thrives on moisture and dirt. To get rid of it, mix one quart household liquid bleach, ⅓ cup powdered detergent, and 2 cups trisodium phosphate (TSP). Make a gallon of this solution by adding warm water to it and use it to scrub down walls and ceilings. After you are done scrubbing with a stiff brush, hose down the entire area thoroughly. Because mildew grows best in shaded areas, trim back all vegetation to let sunlight dry out these areas on the exterior of your building.

SOURCES OF MOISTURE

Moisture to support the process of decay and decomposition originates from several sources: natural moistness found in unseasoned wood, moisture from the ground, rain, and melting snow, condensation, and plumbing leaks. All of these individually or in some combination will cause rot to develop in a building.

By using unseasoned wood in any kind of installation, you take the risk from the start that it will decay and lose some of its original shape by warping and splitting. In turn, it won't be able to hold fasteners (screws, nails, bolts), and your project will have

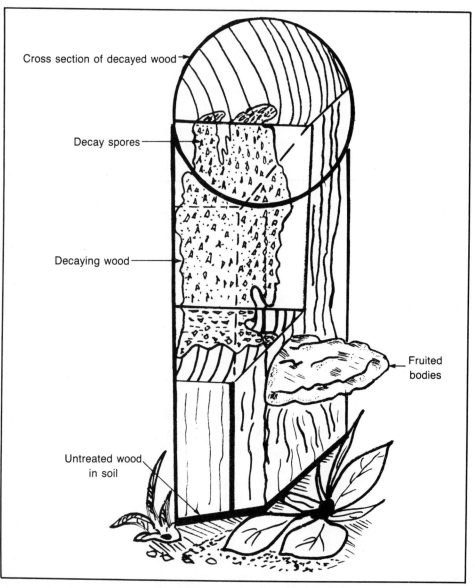

Fig. 5-3. The cycle that produces wood decay. Remember that dry, well-painted, or stained wood will resist decay and last longer than anyone can anticipate.

a good chance of failure. Always buy wood that has been kiln-dried, which most lumberyards sell. Any lumber that you purchase should be stored in a dry area until you are ready to use it.

Ground moisture can get to wood parts of a building in four ways: by direct movement upon contact with soil; by condensation in crawl spaces; by being transported by fungi and by indirect transfer from the soil through concrete or masonry, such as brick or block. Figure 5-4 shows how ground moisture can get to wood.

Fig. 5-4. Wood in direct contact with soil will rot out in time unless it is pressure-treated wood. The photograph shows how moisture can be drawn to the wood and how easy it would be for wood-boring insects to start their trek.

Rain and melting snow cause most of the wood decay in buildings and homes. Decomposition will start at the siding and trim, and water will eventually affect wall studs and sheathing. Of particular concern, always, are open joints and defective flashing areas. In Fig. 5-5 you can see the areas where water has caused damage in this building.

If warm, humid air contacts cold surfaces; it can no longer hold all of its moisture and droplets of water will form on surrounding surfaces. This process, called *condensation*, can occur in most buildings; when it does, the results might not show up for years. Figure 5-6 shows condensation on a water pipe. Find out what to do about excess moisture in your building by reviewing Chapter 3.

To a lesser extent, plumbing fixtures that leak also cause decay in a building. Figure 5-7 shows typical decay found on framing directly under a bathroom. Most leaks usually are detected in time, so serious damage is prevented. Make it part of your maintenance schedule to look for this type of damage.

WEATHERING

Wood that is exposed to the elements is inevitably subjected to surface damage. It becomes a never ending cycle of ultraviolet rays of the sun, water from rain or snow, and a freezing, thawing, and drying out process that will, over a period of years, cause substantial damage to the affected wood areas. Eventually the wood becomes weaker;

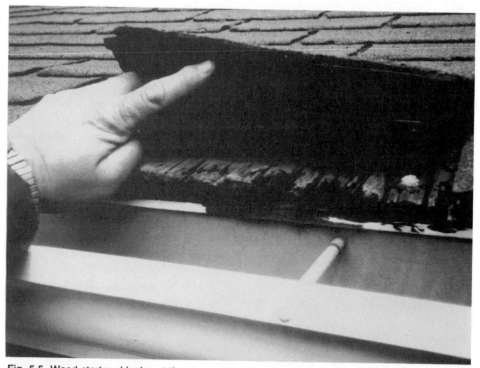

Fig. 5-5. Wood starter shingles at the eaves tend to rot out over the years. When reroofing, insist on replacing this starter edge.

Fig. 5-6. Moisture from condensation can be an available source of water for wood decay to get a foothold in your building.

Fig. 5-7. Wood framing under and adjacent to plumbing fixtures will decay if leaks are not repaired.

it will absorb water and biological agents will thrive in and on it. The absolute best protection for wood against this onslaught is to keep the wood covered with a good quality paint or stain. Table 5-1 gives you some tips on using paint or stain.

CHEMICAL DAMAGE

Wood is damaged chemically by corrosion from rusting metal fasteners such as nails and screws. Figure 5-8 shows what happens when you don't use galvanized or aluminum fasteners. Not only does the wood become discolored from the staining but a process of degeneration in the wood also begins from the chemical reaction of the corrosive materials in the metal. If your building has this problem, then you have your work cut out for you. You should sand, prime, and repaint each rusting area with a rust-inhibiting paint primer. After this is done, re-paint to match the color of the building.

MAINTENANCE AND PREVENTION

Now, with the theoretical subjects out of the way, let's get into the real world. Call it what you want—decay, rot, degeneration, deterioration—it all adds up to the same thing. There is a problem that must be corrected. When wood gets wet and stays wet for any length of time, it is going to change. The change will occur in its color, strength, its physical appearance and characteristics. Changes become noticeable when tight joints

Table 5-1. Tips for Painting and Staining.

✔ For oil-based paints, use brushes with natural bristles.
✔ For non-oil based paints (latex), use brushes with nylon bristles.
✔ Ask your paint dealer to help you select the most suitable paint for your building or paint project.
✔ Be sure to read the manufacturer's directions before applying it.
✔ Always prime new work or bare spots before you paint.
✔ Fill in nail holes and open joints, prime, and then paint.
✔ Prime knotholes with shellac or a primer before painting.
✔ Follow the ladder rules given in Chapter 4.
✔ Never place ladders in front of doors that are unlocked.
✔ Never leave cans of paint on ladders that are not in use.
✔ Keep all paint and stains out of reach of children and pets.
✔ When working indoors, be sure to ventilate the areas to avoid toxic fumes and fire hazards.

Fig. 5-8. Using incorrect types of nails that will rust can cause staining and eventual deterioration in siding and trim areas. The photograph shows the beginning of decay at the rusting nailhead.

open up, when flush wood siding starts to buckle out, and when previously sound wood shows signs of age and decay.

How do you go about preserving your expensive homes and investments against the ravages of decay? The best prescription appears to be: *keep it dry!* Quite often buildings are built without regard to longevity of building materials. Your building might

**Table 5-2. Common Construction
Design Errors and Maintenance Failures.**

✓ Failure to use pressure-treated wood in high risk areas.
✓ Use of defective or decayed wood.
✓ Embedding wood in masonry.
✓ Wood siding or trim in direct contact with the soil.
✓ Wood such as stumps, grade stakes, or concrete forms left in the soil.
✓ Poor or no roof overhang.
✓ Missing or defective roof drainage.
✓ Deteriorated caulking in joints.
✓ Inadequate flashing at windows, doors, and roof areas.
✓ Insufficient or defective vapor barriers.
✓ Minimal or no building ventilation.
✓ Plumbing leaks around and under toilets, tubs, and sinks.
✓ Open joints in building trim and siding.
✓ Poor or no foundation drainage.
✓ Roof leaks.
✓ Improper venting of appliances.
✓ Inadequate stain or paint maintenance.
✓ Flower planters against siding and foundations.
✓ Shrubs and vegetation too close to the building.
✓ Failure to inspect crawl spaces.

have wood siding and trim in direct contact with the soil or embedded in masonry, both of which will invite rot. Of course, failure to maintain your building will also have the same effect. Study Table 5-2 carefully. It is a good reminder of what you should be looking for during your maintenance tours.

The best prescription to keep wood dry indoors is to allow air to circulate at all times. If your building has poor ventilation, the chances of condensation are very high, and condensation spells trouble. Attics and crawl spaces should have sufficient ventilation to prevent moisture from building up and causing problems with the framing members. Review the safeguards for crawl spaces found in Table 5-3 to make sure that moisture doesn't get a foothold on framing members in your building.

TREATED WOOD

One of the biggest boons high technology has given the building industry is the development of pressure-treated wood, which can be purchased at almost any lumberyard. You can distinquish it from regular lumber by its greenish tinge—the result of the chemical treatment it received to make it last for a long time. (Most manufacturers will guarantee this wood for 30 years against rot and wood-boring insect damage.)

Pressure-treated wood is probably the best material to use for outdoor projects, such as wood decks and patio framing. It is still surprising, though, how many contractors cut costs by using untreated wood on exterior projects. It is a known fact that most porches and decks built with untreated wood will start to show major decay in several

Table 5-3. Safeguards for Crawl Spaces.

✓ Provide sufficient ventilation.
✓ Have an easy access for periodic maintenance and inspections.
✓ Add a vapor barrier over exposed soil.
✓ Insulate heat and water pipes as well as heating ducts.
✓ Insulate areas under heated floors.
✓ Remove debris and wood scraps.
✓ Regrade soil away from wood framing.
✓ Keep surface water away.
✓ Inspect for wood-boring insect and decay.
✓ Check for structural defects.

Fig. 5-9. Never use second-hand lumber or wood that is not pressure-treated for outdoor projects. Saving a few dollars cost this owner dearly in the loss of a grape arbor.

years. If you happen to have a deck or porch that has been built with untreated wood, it would be prudent on your part to paint or stain on an annual basis to keep the decay under control.

Use of untreated or defective wood, particularly in an exposed outdoor location, is an open invitation to disaster and a waste of labor. Look at Figure 5-9 to see what happens when costs are cut and poorer quality materials are used. Second-hand, untreated, used lumber was used in the construction of this collapsed grape arbor. Had it been built with pressure-treated lumber, it would still be standing there in its original beauty.

WOOD IN CONTACT WITH SOIL AND VEGETATION

As has been said so many times, untreated wood in direct contact with the soil will rot out and even become home to a variety of wood-boring insects. Figure 5-10 is a prime example of wood contact with the soil. Should you have property that has either siding or trim in direct soil-ground contact, regrade the foundation soil to allow several inches of clearance between them.

Wood does not have to be directly in contact with the soil to draw moisture from it. If it is close enough to the soil, it can draw some moisture from it by capillary action. This is not only true of low wood areas but also of crawl space areas. If your property has a crawl space that is accessible, by all means follow the steps outlined in Table 5-3.

One of surest ways to hasten the rotting out of your wood is to plant various types of plants and shrubs too close to siding and trim. Figure 5-11 shows such a folly. Plantings that are too close to the building will hold moisture long after the rain has gone and also will provide too much shade, thereby promoting unwanted decay. Trim back all overgrown vegetation near your property and avoid watering directly against wood siding or trim.

While on the topic of vegetation, never place flower planters next to foundation or siding areas of a building. Even treated wood boxes will leak water to the siding. In Fig-

Fig. 5-10. Always be on the alert to any areas that have direct wood-soil contact. As you can see, the corner support post appears to have rotted out.

113

Fig. 5-11. Keep all of your plantings under control and several feet away from foundation areas as well as wood siding and trim areas.

ure 5-12 you can see the mistake that this owner has committed. This flower planter box will be host to more then just flowers.

PLUMBING LEAKS

Some wood decay can be directly traced to plumbing leaks. Even small leaks can do a job on sound wood framing. Leaks around tubs, sinks, under toilets, and from pipes are common and can be easily repaired. During your inspection-maintenance tours, be on the lookout for deteriorated tile caulking, particularly in wall and floor joints. If caulking looks anything like that as seen in Figure 5-13, then you better get busy sealing up these open gaps.

Caulking is a fairly simple procedure, especially if it is done on a regular basis. There are, however, a few professional tips that you should know about. First, always be sure to scrape out all the loose and deteriorated caulking in the joints. Once you have a clean, open joint, apply a bead of caulking to the gap. Use your finger to smooth the caulking into the joint so that you end up with a nice, tight joint. If you are caulking the tub-wall tile joints, use the following trick to give a long-lasting tight joint. Fill the tub with water (you might even want to get into the tub as if you were going to take a bath). The idea here is to get the tub at its lowest position before you apply the caulking. After the caulk has set, wait a few hours and then drain the tub. Any further movement will not adversely

Fig. 5-12. Do not place wood planter boxes adjacent to the building. Even with treated wood, some moisture will be drawn to the building siding resulting in decay.

Fig. 5-13. Open joints in tile wall and floor areas will let water into the interior areas and, in time, cause decay and deterioration of structural areas. Be sure to caulk and seal all such open joints on an annual basis or as required.

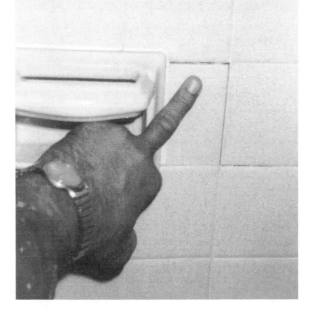

Table 5-4. Common Caulking Compounds.

Type	Use	Comments
Oil based	Routine caulking and sealing	Shrinks; maximum life span is five years; least expensive
Oakum	Wide or deep cracks	Needs to be covered with additional sealing
Butyl-rubber	Metal-to-wood or metal-to-masonry joints	Difficult to use; can be painted over immediately
Neoprene	Cracks in concrete	Toxic fumes
Silicone	Joints that require waterproofing	Long lasting (10 years); costly in comparison

affect the new caulking because the seal was applied at its widest opening. Use a silicone, rubber-base caulking compound and follow the manufacturer's direction.

Caulking joints outside is just as important to do as it is for the inside of the building. Open joints at windows, doors, and flashing areas should be sealed up tight on an annual basis to provide protection against leaks. Table 5-4 lists the more common types of caulking materials that can be used to seal up open joints in your building.

SUMMARY

As you have seen, decay is the result of poor design, faulty construction, and the lack of a full, on-going preventive maintenance program. If wood is kept dry, it might last indefinitely; but once the wet-dry cycle starts, a whole Pandora's box of biological and chemical deterioration is opened up. The crux of keeping wood from rotting away is to keep it dry. You must *always* be on guard—against the many sources of moisture, and you must constantly provide safeguards. Consult Table 5-5 every so often to see where decay is likely to begin and Table 5-6 to find out what you can do to prevent it. If you follow these rather simple guidelines, you might be able to say that your building is relatively decay-free.

STRUCTURAL PESTS

Once the decaying process in wood has begun, its natural progression is hastened by the variety of wood-boring insects that stake out your property for an easy but nevertheless delectable meal. Insects such as termites, powder-post beetles, carpenter ants, and carpenter bees are all crazy about punky, moist, and easy-to-chew wood. But

Table 5-5. Places Where Wood Decay Is Likely to Occur.

- ✓ Base of door jambs
- ✓ Foundation windows (particularly if in contact with soil)
- ✓ Open joints in siding, trim, gutters, and flashing
- ✓ Flat surfaces such as on decks, flooring, railings
- ✓ Base of wood posts
- ✓ Window trim
- ✓ Attic chimney framing
- ✓ Wood gutters (particularly the joints and downspout connections)
- ✓ Gutter trim and adjacent framing
- ✓ Base of stair stringers
- ✓ Low siding and trim
- ✓ Wood in contact with soil
- ✓ Base of garage doors
- ✓ Wood skylights
- ✓ Wood adjacent to defective flashing
- ✓ Wood that is constantly exposed to wettings
- ✓ Wood under and adjacent to leaking plumbing

Table 5-6. Measures to Prevent Wood Decay.

- ✓ Use pressure-treated wood for exterior use.
- ✓ Keep wood dry.
- ✓ Repair plumbing leaks.
- ✓ Caulk and seal all open joints.
- ✓ Never use untreated wood in direct contact with the soil.
- ✓ Repair defective flashing.
- ✓ Provide and maintain an efficient roof drainage system.
- ✓ Provide sufficient ventilation in attics and crawl spaces.
- ✓ Vent all appliances to the outside.
- ✓ Regrade foundation soil to allow water to flow away from the building.
- ✓ Regrade soil to provide several inches of space from low wood.
- ✓ Remove nonfunctional wood from soil, e.g., stumps, grade stakes, etc.
- ✓ Provide vapor barriers in crawl spaces.
- ✓ Keep all exterior wood painted or stained.
- ✓ Trim back all vegetation too close to the building.
- ✓ Avoid putting planters next to the building.

that is not the only kind of wood that they will consume. Wood-boring insects are known to chew dry, seasoned wood as well. So, even though your building might show no signs of decay, you do have to be constantly on the "lookout" for the activity of structural pests.

TERMITES

Termites are the most destructive of the structural pests. They eat with a ferocious appetite and cost the American public millions of dollars each year, not only in repairs

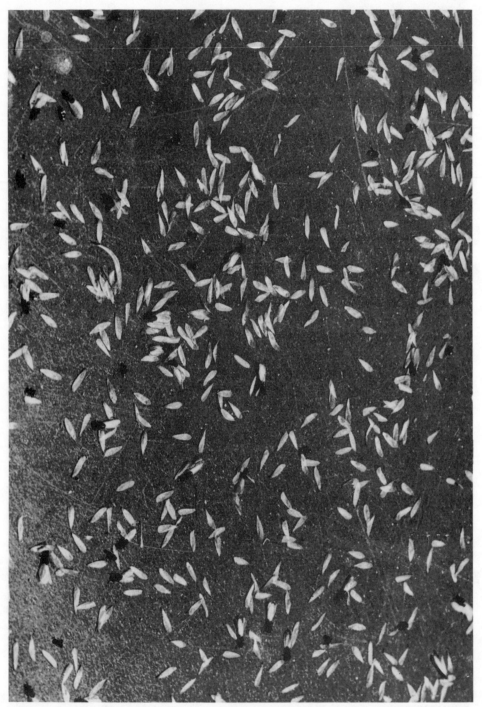

Fig. 5-14. Discarded wings indicate an active termite or carpenter ant colony in your building. (Courtesy of U.S. Forest Service)

but also for attempts to control their numbers. Because knowledge of nesting and feeding habits can only help you save money and grief, be sure to read the text very carefully. Also, for a very detailed account, read Chapter 7 in *What's It Worth? A Home Inspection and Appraisal Manual*. This book is listed in Appendix B.

Reproductive or swarming termites are the young males and females of termite colonies that are sent out from their colonies to reproduce and start new colonies. Sometimes people will confuse them with swarming ants (which are the reproductives of the carpenter ant colonies). Here are three ways to distinquish between the two:

- *By the shape of the body.* The ant has a narrow wasp-waist; the termite is straight down with no narrowness at the waist.
- *By the wings.* The ant has two pairs of wings of which the front pair are much longer than the rear pair; termites, on the other hand, have two pairs of wings of equal length.
- *By the antenna.* The ants' antennas bend at right angles and, unlike the termite, do not have any bead-like configurations on the antennas; the termites' antennas are straight with no bends and have bead-like configurations on it.

If you find either species in your building, you can bet that you have an established colony nesting somewhere.

How to Recognize the Presence and Work of Termites

Large numbers of winged reproductive termites that emerge or swarm from soil or wood are likely the first signs that a termite colony exists. Discarded wings, as seen in Figure 5-14, also point to the existence of a well-established colony nearby. If you see such wings beneath doors and windows, be alert because termites might have tried to escape from within the building.

The presence of mud tunnels (Figure 5-15) is another sign of termite infestations. These tunnels range from ¼ to ½ inch or more wide. Termites used them as passageways between wood and the soil from which they draw their moisture. The tunnels protect them from the drying out effect of direct exposure to the air.

Termite-damaged wood is not noticeable on the surface. Termites are quite skilled at hollowing out wood without damaging the outer surface. The only way to determine whether termites are presently active or have been active in a building is by probing with a tool such as a screwdriver, as seen in Figure 5-16. If the wood appears hollow and the tool penetrated easily, you can assume that termites have been at work. If in addition you find what appears to be tiny white ants, then you know that not only have they been busy but they have never left. Get an exterminator in there to treat the house.

Preventing Termite Attack

To make the environment as inhospitable as possible for any kind of wood-boring insects, remove all wood scrapes, roots, stumps, and other wood debris from the soil around your building. Also make sure that moisture doesn't get a chance to build up around your property by paying attention to proper soil grading and roof drainage runoff. Check your foundation for cracks and deterioration. Seal up any openings that could allow termites easy access. Cracks as small as ¹⁄₃₂ inch in width will permit termites to "wander" in. Walls of concrete block should have the top courses capped off with cement.

Fig. 5-15. The presence of mud tunnels shows that termites have been clever enough to devise a way to continue their march through your building.

Unsealed openings in these foundations are always an open invitation for guests you don't want.

Termite Maintenance Program

Exterior
- Check foundation walls for mud tunnels.
- Inspect behind overgrown shrubs and bushes for evidence of termites.
- Look for cracks and open joints in foundation walls.
- Probe all low wood areas for signs of termite damage.
- Inspect crawl spaces for mud tunnels and discarded wings.
- Pay particular attention to untreated wood directly in the soil.

Interior
- Check inside foundation walls for mud tunnels.
- Probe all low wood areas.
- Look for discarded wings.
- Test the condition of foundation framing members such as sills.
- Make sure that hollow core foundation blocks are sealed at the top.
- Repair any cracks you find in your foundation walls.

Eliminating Termites

If you find evidence of active termites during your maintenance tour, consult with a local exterminator about how to get rid of them. Once the chemical treatment is done

Fig. 5-16. Use a tool such as a screwdriver to probe suspect areas during your quest for termites. If you find hollowed out wood, you have had these unwanted guests, and they might still be lurking about.

and you have made the necessary repairs and alterations in your building, you should be reasonably free of termites. It would be wise to have the exterminator come back on an annual basis to recheck for the possibility of a recurrence of these insects, which then would require treatment again.

BEETLES, ANTS, AND BEES

Termites are not the only insects that make their living by chewing up your building. Powder-post beetles and carpenter ants as well as carpenter bees can also cause serious damage if they are left unchecked. The kind of damage done by these insects differs from that done by termites. Powder-post beetles leave a fine-to-coarse wood powder that is packed tightly in galleries of the damaged wood. Carpenter ants, on the other hand, build hollow, irregular, but clean chambers across the grain of the wood. Carpenter bees bore into the wood and leave entry holes that are almost perfectly round and approximately ½ inch in diameter.

Powder-Post Beetles

Again, during your maintenance inspection for wood-boring insects, look for the obvious signs that powder-post beetles leave behind—small round emergence holes in

Fig. 5-17. Wood damaged by powder-post beetles.

the infested wood. These emergence flight holes are approximately $\frac{1}{16}$ to $\frac{1}{8}$ inch in diameter. Figure 5-17 shows such an example. Shine a flashlight on the holes to determine whether the activity is old or new. Newly formed holes have a light, clean appearance whereas older ones are much darker in color. To gauge the extent of the damage, take a screwdriver and probe the wood. If it looks anything like what you see in Figure 5-18, you should consult with a structural engineer about the soundness of your building. If you are reassured by the engineer that your building is sound, than you will want to follow some simple steps to keep it that way. Keep all untreated wood surfaces painted and bear in mind that any repairs or replacements of wood should only be done with pressure-treated wood. As with any preventive suggestions about wood, keep wood framing free from the attack of moisture.

Carpenter Ants

Irregular, hollow, clean chambers cut across the grain of the wood; fine-to-course wood fibers discarded by the ants at nest building time; numerous black to reddish brown looking ants, $\frac{1}{4}$ to $\frac{3}{8}$ of an inch long, roaming around the inside or outdoors; flying ants and discarded wings near openings such as window and doors—all are signs that carpenter ants are present. So during your maintenance-inspection tours, be sure to check areas that easily succumb to decay, especially wood gutters, low wood areas such as foundation windows, untreated wood directly in the soil such as landscape timbers, porches and

Fig. 5-18. Damage of the magnitude illustrated in this photograph requires the expert opinion of a structural engineer as to whether the building is still sound. (Courtesy of U.S. Forest Service)

spaces, and any areas that are normally hidden from view such as behind shrubs and bushes.

Carpenter ants will nest almost anywhere in a building as long as they can find wood that is rotting away or has been softened up by moisture. Hollow-core doors in buildings are favorite nesting sites for ants. As should be done with any rotting wood, remove it and replace it with pressure-treated wood. If you find evidence of ants on your property, treat with an approved insecticide such as diazinon. Figure 5-19 shows what happens when owners fail to carry out a preventive maintenance program.

Carpenter Bees

Carpenter bees are wood-boring insects that look like large bumble bees with the exception that they do not have the typical bumble bee fuzz covering their abdomen. After mating, the female carpenter bee will look for a suitable nesting site, such as an unpainted piece of exterior wood surface. Once she has made her decision, she will excavate a shallow entrance hole approximately ½ inch in diameter and then tunnel laterally for a few inches more. It is in this right-angle tunnel that she will lay her eggs. After the eggs have hatched, the emerging new bees mate and start the cycle all over again.

Fig. 5-19. Carpenter ant damage in a a foundation window sill.

The continued drilling into the wood could cause major damage. Don't give these bees the chance to damage your building. An exterminator can easily spray and control these insects. You can also help out by making sure that your building is well painted and that no bare wood is left to be used as a nursery for carpenter bees.

SUMMARY

As you have read, the natural progressive step in hastening the demise of the framing in your building is carried out by wood-boring structural pests. These insects that do all of the dirty work to your building can be easily identified, however, by both their characteristics and by the type of damage that each species leaves behind. Check Table 5-7 to review the kinds and types of damage as well as the control measures. In Table 5-8, you will review steps that will keep your building virtually free from wood-boring insects. For a very helpful booklet on controlling decay and insects read, *Finding and Keeping a Healthy House*, which is listed in Appendix B.

Table 5-7. Wood-boring Insects.

Type	Damage They Cause	Evidence of Activity	Control Measures
Termites	Hollow out wood and cause major structural damage.	Discarded wings; mud tunnels; swarmers—reproductives; damaged wood.	Remove all wood from direct soil contact; treat chemically.
Powder-Post Beetles	Reduce wood to a powdery substance; could cause major damage.	Flight holes; powdery wood.	Keep wood dry and painted; treat chemically.
Carpenter Ants	Moderate damage if caught in time; damage wood by nesting in it.	Hollowed out nests; sawdust near nests; discarded wings; swarming ants.	Keep wood dry; use pressure-treated wood; treat with diazinon spray or dust.
Carpenter Bees	Minor damage unless left unchecked.	½-inch clean, round entrance holes in wood siding or trim.	Keep exterior wood painted; have an exterminator treat with a pesticide.

Table 5-8. How to Prevent Wood-boring Insect Infestations and Damage.

✔ Keep wood dry.
✔ Paint and stain on a regular basis.
✔ Remove and replace decay and/or deteriorated wood.
✔ Where practical, replace damaged wood with pressure-treated wood.
✔ Never put untreated wood in direct contact with the soil.
✔ Never bury wood in the soil near your building.
✔ Provide proper roof drainage away from your building.
✔ Provide sufficient ventilation in basement, attic, and crawl spaces.
✔ Keep all exterior joints (siding and trim) caulked and sealed.
✔ Inspect crawl spaces every six months.
✔ Remove dead and dying trees near your building.
✔ Remove vegetation too close to your building.
✔ Remove tree stumps.
✔ Grade all soil away from low wood areas.
✔ Have an exterminator inspect—on an annual basis—buildings constructed with hollow-core concrete block, brick, stone, or slab construction.
✔ Call in a professional if you suspect problems.
✔ Follow the manufacturer's directions in the application of pesticides.
✔ Keep up your inspection-maintenance program.

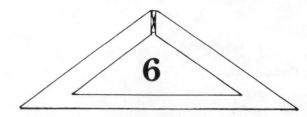

Energy Maintenance

This chapter will describe an energy maintenance program for your building. It will teach you energy maintenance techniques that could save you hundreds and maybe thousands of dollars a year. You should be warned, however, that unlike the many books being published today, you are not going to learn about a lot of exotic or even expensive ways to cut back on your energy dollars. As you read on in this chapter, you will find many useful, inexpensive, but practical suggestions that you can easily implement in your energy maintenance program.

ACTION-REACTION

Keep in mind that for every action there is a reaction—and this certainly holds true for energy savings. If a building is "too tight," you will clearly save money on fuel costs, but conversely, you might end up spending hundreds of dollars in repairs for decaying wood and peeling paint. Figure 6-1 shows the results of a tight house where condensation-moisture problems are quite evident. As you learned in Chapter 3, making a house or building too tight will cause a host of problems, including indoor pollution which will be covered in Chapter 8. With this in mind, let the word "moderation" be your byword.

KISS PRINCIPLE

If you have read the book *What's It Worth*, you will know that it is best to adhere to the Kiss Principle, which is to *keep it strictly simple*. The following energy conservation hints are based on low or no-cost approaches that you can do yourself to save energy and money in the running of your property. Some of the ideas are as simple as lowering

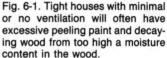

Fig. 6-1. Tight houses with minimal or no ventilation will often have excessive peeling paint and decaying wood from too high a moisture content in the wood.

your thermostat setting, while others require a small cash outlay, such as for caulking compound. Virtually all of them will pay off "big" for you in the long run.

HEATING AND COOLING SYSTEMS

Expect big paybacks with some small adjustments that you can make to your heating or cooling systems. For example, changing your thermostat setting will cost you nothing, but you can save hundreds of dollars each year by doing it. Keeping your burner tuned up on a regular basis will ensure that it will perform better and will last longer because repairs will be cut to a minimum. Table 6-1 lists some very good, simple ideas that you might want to include in your maintenance program. A more detailed analysis of mechanical systems will be covered in Chapter 7.

REDUCING AIR INFILTRATION

Air infiltration is one of the reasons why your bankbook balance is always so low. The invisible transfer of warm and cold air into and out of buildings goes on continually, stealing money right out of your pocket. This exchange of air is possible because your building can be full of open gaps and cracks to which you have paid little if any attention. You should, though, because sealing up these openings can help reduce your heating and cooling costs. Let's take a look at some easy ways to cut back on these air losses.

Table 6-1. Heating and Cooling Energy Conservation.

✔ Lower or raise thermostat settings for heating or cooling.
✔ Consider having an automatic setback thermostat installed.
✔ Wear more clothing in winter and less in summer.
✔ Follow manufacturer's recommendations on thermostat setbacks for heat pumps.
✔ Bleed hot-water radiators seasonally.
✔ Change filters on warm air systems as needed.
✔ Replace defective fan belts and adjust loose ones in warm air systems.
✔ Vacuum warm air registers and ducts as required.
✔ Seal up open joints in heat ducts.
✔ Replace defective air vents on steam radiators.
✔ Don't block the free flow of air around radiators or registers.
✔ Vacuum dust from radiators and heat fins.
✔ Insulate heat pipes and ducts.
✔ Have burners tuned up annually.
✔ Be sure to have a professional inspect and tune up air-conditioning systems.
✔ Replace or clean filters in room window air conditioners.
✔ Vacuum evaporator and condensor coils on air conditioners.
✔ Only buy appliances with high EER ratings.
✔ Always follow the manufacturer's recommendations for service.
✔ Don't waste valuable energy resources.

Caulking

Check out the original caulking that sealed your storm windows. Does it need replacement? If so, do take the trouble to recaulk not only the top of the storm windows but also the sides. Use an acrylic-latex or latex caulk for best results. Make sure also that the storm windows have weep holes in them, as seen in Fig. 6-2, to give excess moisture a way out. Do not caulk over these important areas of your storm windows.

If you have large gaps that have to be sealed, consider using rope caulking. It can be purchased in almost any building supply store. Rope caulking is a flexible, putty-like material that can be installed in any opening with ease. One nice thing about it is that it can be used over again if you need to remove it.

Windows are not the only places that need caulking. Figure 6-3 shows you the most common caulking areas of a typical building. Refer to this illustration when in doubt about where to caulk. Places such as chimney flashing joints, foundation sills, plumbing openings, and anyplace where there is an opening in a building wall, foundation, or roof will need to be inspected and checked for air leaks.

Weatherstripping

Make sure that you have included in your maintenance program periodic inspections of the weatherstripping on doors, garage doors, and attic and crawl-space access hatches.

Fig. 6-2. Open up or provide weep holes in all storm windows to allow trapped moisture to escape before it can do any damage.

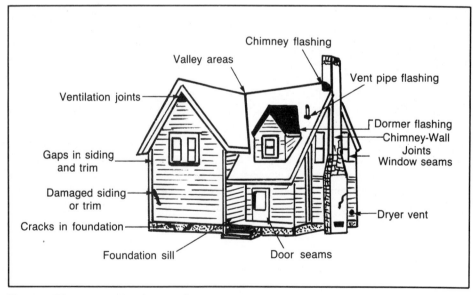

Fig. 6-3. Where to caulk and seal. Check annually for damage and deteriorated caulking and weatherstripping on all openings.

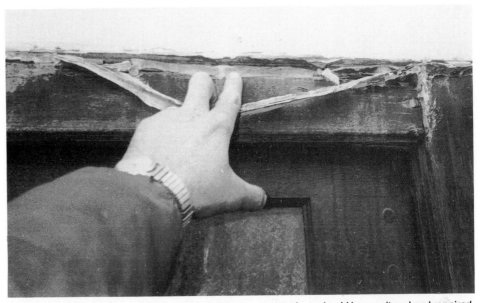

Fig. 6-4. Weatherstripping on doors, in this instance a garage door, should be monitored and repaired on an annual basis.

Of particular importance are hatches to unheated areas. All the insulation in the world in your attic or crawl space won't do you any good if you have air leaks around these openings. If you find that you either have no weatherstripping or that your weatherstripping has deteriorated, be sure to repair or replace it. In Fig. 6-4 you can see how the weatherstripping on this garage door is no longer doing its job. Make it a maintenance point to check before the winter season for any weakness in your line of defense. Every six months go around to these areas and reseal, nail down, and generally tighten up these potential air leak areas.

INSULATION

Because volumes and volumes have been written about all the different types of insulation and how and where to install it, this chapter will only provide you with the important essentials. For additional reading on insulation, see Appendix B.

The best and first place that you should consider insulating is your attic floors. Probably the worst place to insulate would be under the attic roof framing, particularly if no consideration is given for ventilation. If you have a typical attic, one which is open and used primarily for storage or just dead space, make sure that there is plenty of insulation on the floor between the floor joists, as seen in Fig. 6-5. Depending upon the area of the country that you live in and the type of heating system that you have, a minimum of 6 to 12 inches should be sufficient. Check your attic, and if you find that you are under the figures given, then by all means add several inches.

Insulation that is applied to the underside of attic roofs should be installed by a professional. Trapped moisture between the insulation and the roof framing can develop into major damage to the insulation and decay in the wood framing. Special precautions will have to be taken by the professional (roofer or carpenter) to provide a sufficient

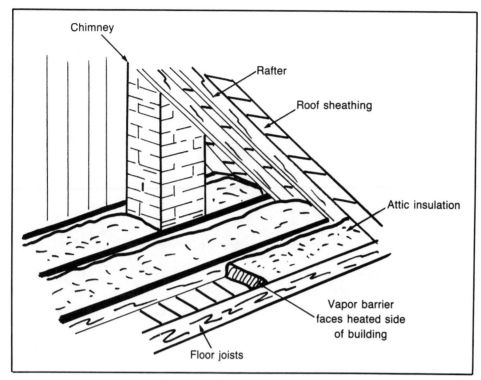

Fig. 6-5. Be sure that your attic and crawl-space floors have sufficient insulation. The few dollars you spend will be paid back the first few winters and summers.

Table 6-2. Insulating and Plugging Up Air Infiltration.

✔ Caulk and seal all open gaps, cracks, openings, and joints.
✔ Plug large holes with insulation or oakum and then caulk.
✔ Use only quality caulking compounds.
✔ Repair damaged weatherstripping.
✔ Install weatherstripping where missing.
✔ Insulate attic floors.
✔ Insulate and weatherstrip attic access doors.
✔ Provide vapor retarders towards the heated sides of buildings.
✔ Insulate foundation sills.
✔ Insulate under the first-floor flooring.
✔ Provide ample ventilation and vapor barriers before insulating.
✔ Think twice before going to the added cost of insulating exterior walls.
✔ Make fireplaces tight by providing glass enclosures.
✔ Install switch and outlet gaskets on exterior wall receptacles.
✔ Cover or remove window air conditioners during the off-season.
✔ Provide sufficient ventilation to avoid condensation.

air gap between the wood and the insulation. In addition, soffit and ridge vents must be installed to vent out the areas where the insulation is installed. Truly, this is a job best left to a "pro."

An easy job that you can do, however, is to insulate the foundation sills (Fig. 6-6). This is definitely a do-it-yourself task and the payback is very good. For a few dollars of insulation and a couple of hours of work, you can make the sill areas tight as a drum.

Crawl spaces are another area that need your attention. Usually the areas over the crawl spaces are heated living areas and, as such, should have the benefit of sufficient insulation under them. Staple several inches of insulation to the underside of the rooms above these spaces. Staple the insulation in between the floor joists with the vapor barrier facing the heated side (towards the living areas). Because of the importance of ventilation and vapor barriers, it will come as no surprise that additional vapor barriers are required in crawl spaces. To keep moisture down, consider laying a plastic vapor barrier on top of the exposed soil in the crawl space. Figure 6-7 shows how to do this. Notice the ample use of vents to air out the crawl space. If you have a crawl space with limited or no ventilation units, consider adding them.

Insulation pumped into the exterior wall cavities is something that you might want to consider. Before you do, however, ask yourself the following questions: are your windows tight, do you have storm windows or thermal windows, have you caulked and sealed to make your building weathertight, do you have sufficient insulation in attic and crawl-space areas, and is your weatherstripping intact and in place on your building? If

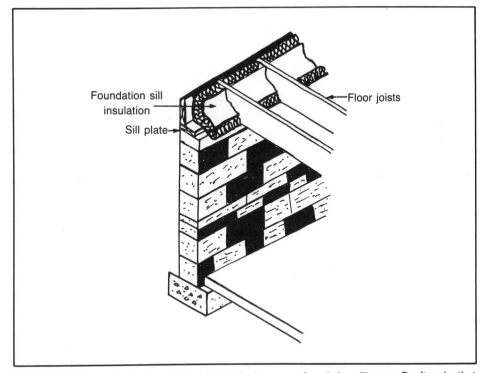

Fig. 6-6. An often overlooked area to insulate is the basement foundation sill areas. Don't make that energy loss mistake.

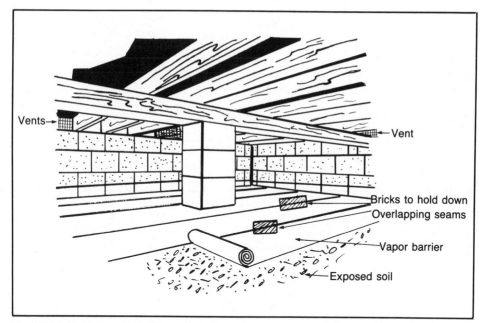

Fig. 6-7. Vapor barriers (sometimes referred to as vapor retarders) should always be placed over exposed soil to keep down any vapors that could result in condensation.

the answer is no to any one of these questions, hold off the expense of blown-in insulation until you take care of the unfinished easier projects. With any air leaks in your building, the blown-in insulation would be a waste of money.

OTHER AIR LEAKS

Other areas that allow air to escape or infiltrate are fireplaces, exterior wall switches and outlets, and window air conditioners. One of the easiest ways of cutting back on fireplace heat losses is to install a glass enclosure over the fireplace opening (Fig. 6-8). Make sure that it is a tight-fitting unit and that the edges are sealed and caulked. Stop drafts that come in through electrical receptacles by installing styrofoam or foam rubber gaskets that fit behind the plate. Gaskets are inexpensive and can be obtained at most hardware stores. Always make sure that electricity is off when you are working on any electrical outlet or switch. Remove window air conditioners during the off-heating season. Cover window units that are too large to remove with an air-conditioner cover or a sheet of plastic to keep out the weather.

VENTILATION

Insulation and tightening up your building to save energy and costly fuel bills is very commendable, but you have to watch out for the old "double-whammy," which is the action-reaction syndrome mentioned earlier. If you make your building too tight without letting it breathe, you will be, in effect, killing it. Your building will huff and puff but it won't have anyplace to exhale all of the moisture that accumulates. (See Chapter 3 which

134

Fig. 6-8. Fireplace enclosures need to be caulked and sealed to keep out unwanted drafts and to avoid heat losses from the building.

deals with condensation.) The end result of all of this could be decay and deterioration in your building. Because you don't want that, pay particular attention to the following information.

Location, location, location—the three most important factors in buying real estate—can now be changed to ventilation, ventilation, ventilation—the three most important factors for a healthy house. You bought your building based on location, and now you must keep its worth by providing sufficient ventilation.

Read this next statement and ingrain it in your mind. *You can never overventilate a building, as long as it is sufficiently and properly insulated.* In other words, always remember that ventilation is one of the keys to happy and successful ownership of your property. Underventilate or provide no ventilation to your building, and you might as well send out written invitations to both decay organisms and wood-boring insects. Figure 6-9 shows a building that probably has internal sweating problems.

In Table 6-3 you will find some ventilation tips about types of vent units with a rating system. This table runs the gamut from the worst to the best possible combinations. Use it to gauge how well your building is ventilated. If your building falls into any of the categories found in Table 6-4, be especially careful about checking out your building for possible problems.

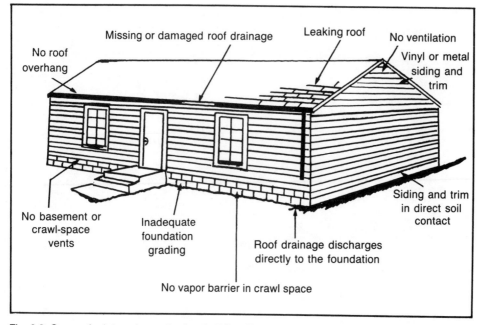

Fig. 6-9. Causes for internal sweating in a building. If your building has such areas, be sure to make corrective repairs.

Buildings That Require Extra Ventilation

Some types of buildings require more than the average amount of ventilation. This extra ventilation is necessary because of construction and design of the building; for example, if a building has aluminum or vinyl siding, moisture often gets trapped between the siding and the building framing. Although you might not see it, the wood begins to rot and insects settle in to do their thing. Unlike wood products, these synthetic materials cannot breathe. Moisture from condensation is almost guaranteed. To avoid this, you must provide additional ventilation to air out the building.

Warm-air heating systems—especially when combined with a built-in humidifier—generate a great deal of moisture that permeates the air within the building. Not only does the moist air in the building cause problems as described, but also the humidifiers in the heating systems in time tend to cause serious corrosion problems within the heating plant. This should give you some food for thought if you own such a heating plant. (Chapter 7 will discuss heating system problems in depth.)

Buildings built over crawl spaces, in higher water table areas, and with blown-in insulation in the exterior cavity walls also will show some signs of moisture/condensation problems. Such buildings also require additional ventilation.

Types of Vents

You have several options when it comes to choosing types of ventilation units. Figure 6-10 illustrates some of the more common types.

The *gable vent* is one of the most popular type of vent. Like all vents, it must be installed as high up on the building as possible. It also must be complemented by opposing

136

Table 6-3. Ventilation Tips.

Type	Rating	Comments
None	Extremely poor	Expect condensation; decay, and wood-boring insects
Drip-edge vents	Very poor	Vent openings too small; will need additional vents
Motorized fans	Fair to poor	Costly to run; might not operate in winter if thermally activated
Turbine vents	Fair	Depends upon wind movement to activate
Ridge vent alone	Poor	No convection of air
Soffit vent alone	Poor	No convection of air
Roof-mounted vents alone	Poor	Little convection of air
Siding wall vents	Poor	Clog up with paint and insects
Crawl space foundation vents	Good	Need several of them for good cross ventilation
Gable vents	Good	Need to be placed opposite each other up high on gable ends
Combination ridge and soffit vents	Good	Watch out for obstructions such as insulation
Combination of ridge, soffit, and gable end vents	Excellent	Probably one of the best combinations for effective attic ventilation

vents to allow good cross ventilation. Just having one vent or a number of vents all on the same side is almost as bad as having none.

Watch out for ventilation units that require either the wind to move them, such as *turbine vents* (Fig. 6-11), or *motorized units* that must be thermally activated or turned on manually. The wind is not always at your beck and call, and motorized units run on costly electricity, which is what you are trying to save (money) and not waste.

Combination ridge and soffit vents (Fig. 6-12) are an effective way to ventilate an attic. Keep in mind that having one without the other defeats the purpose, namely the movement of air. Probably the best combination would be the *soffit and ridge vent in combination with gable vents*. This would give you the best of both worlds.

**Table 6-4. Building
Features that Require Extra Ventilation.**

Type	Comments
Metal or vinyl siding	Tend to hold and trap moisture
Warm air heating systems	Pump a lot of moisture throughout the building
Humidifiers in heating systems or being used as room units	Provide extra moisture to the air in buildings
Buildings built over crawl spaces	Need vapor barriers and foundation vents
Buildings built in high-water table areas	Need to ventilate basement and crawl spaces more effectively
Buildings with blown-in insulation	Tend to trap moisture in the wall cavities, particularly if no vapor barriers were provided

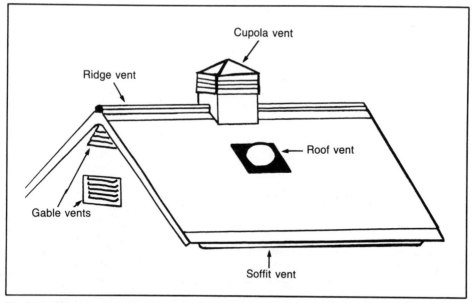

Fig. 6-10. Several options for roof vents. Remember that air movement needs more than one vent; usually a combination of vents is best.

Roof-mounted vents (Fig. 6-13) are suitable for most roofs, particularly hip-style roofs. If your building is this style, don't skimp; put several of these units on your roof. Remember to have them installed at the highest point possible and preferably position them on opposite sides of the roof for a better flow of air.

Fig. 6-11. A roof-mounted turbine vent. Vents that require wind movement or electrical means to vent out air are probably not the most efficient or effective. Natural air convection has proven, when used in combinations, to be the most effective and cost efficient.

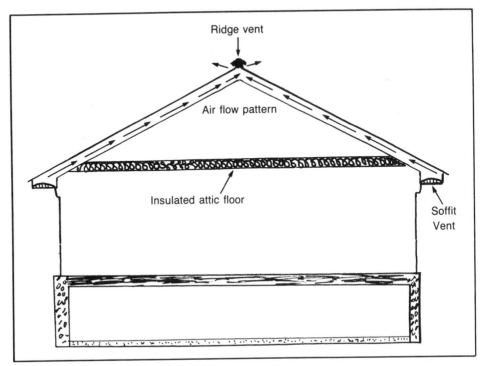

Fig. 6-12. Combination vent units work best.

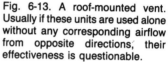

Fig. 6-13. A roof-mounted vent. Usually if these units are used alone without any corresponding airflow from opposite directions, their effectiveness is questionable.

A relatively new addition to the ventilation marketplace is the *drip-edge vent*, as seen in Fig. 6-14. Like so many newer things that you can buy these days, it has some questionable considerations. Because the openings in the vent are small and it can only be installed at the drip edge of the eaves, it provides very little ventilation to the building. If you have a choice, stay away from these.

What, then, is the most suitable ventilation units for your building? The answer is— stay with a combination of ridge, soffit, and gable vents, as large as your roof possibly allows. As long as you get a natural convection of air flow patterns that carry moisture and heat out of your attic without help from costly electricity or unpredictable winds, you should be all set. One important last thing to remember, though: These vents must be left open year around. Don't close them off for the winter thinking that you are losing valuable heat. In a well-insulated building, you don't have to worry because the vents ventilate out only unwanted moisture and excess summer heat.

Fig. 6-14. Once more new technology does not prove to be 100 percent effective. The new drip-edge vents found both on new construction and retrofit work leave much to be desired as far as volume of air movement provided by these vents.

WINDOWS

Windows are both good and bad as far as energy conservation is concerned. Good in that they allow in much needed sunlight to warm your building free of charge, but bad in that windows are one of the worst offenders when it comes to heat losses. Let's check out a few ways to help make windows less of a culprit.

During your inspection-maintenance program, note any loose windows, deteriorated putty or caulking, damaged parting beads, broken sash cords, and certainly any cracked window panes. Do the repair work as soon as you can, even if it is just to temporarily place a piece of clear tape over a crack in the glass. And here is a helpful hint that might come as a surprise to you. Dirty glass can block up to 40% of the solar energy that comes through during the day. So dirt can affect your heating bill by as much as 3 or 4%. Put a little elbow grease to work and get those windows sparkling clean!

As soon as the heating season starts, close all of your storm windows and lock the inside sashes. If you have storm windows, the middle pane of glass belongs at the bottom

Fig. 6-15. Window hardware such as the clamshell latch provide a tight seal between the upper and lower windows. If you have windows without such locking devices, be sure to add the chore of installing them to your maintenance list.

and the other one at the top. They seal best in this position, so don't make the mistake of reversing them and leaving an open gap. Double-hung windows—and this is very important—should be locked together with the clamshell latch (Fig. 6-15). If you find any windows without latches, pick some up at your local hardware store and install them.

OTHER HELPFUL HINTS

Have you ever thought about the heat that you lose from radiators because of heat transfer to the outside walls behind them? Homemade radiator reflectors are one inexpensive way to curb such losses. You can either have them made of sheet metal, or you can make them yourself with aluminum foil cemented to a thin layer of rigid insulation material. The radiated heat strikes the reflector that, in turn, bounces the heat back into the room where you want it in the first place. Figure 6-16 shows an example of this.

Some experts will tell you that the exhaust from your clothes dryer is a good source of cheap heat. Dryer vent diverters are sold in supply stores and appliance stores. Figure 6-17 demonstrates this type of heat reclaimer. My comment is—*don't use them*! Too much moisture will be thrown back into the basement or building proper; furthermore, you might vent dangerous combustion by-products into your home via this method. Better pass on this one.

Keep your appliances in good working order and you will save money. Periodically, clean and check them all out, particularly the heavy energy users. See Chapter 9 for a complete rundown on maintenance tips, including specific tips for appliances.

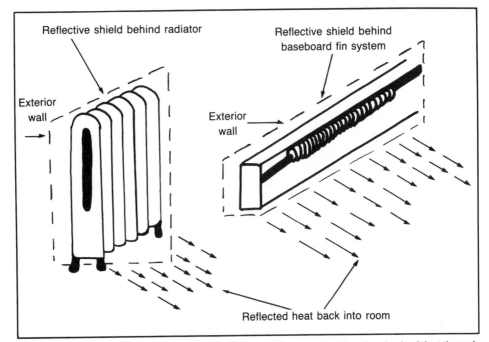

Fig. 6-16. A simple cost-effective way to radiate heat back into a room rather than having it lost through exterior walls.

Fig. 6-17. Heat reclaimers can do more harm than good. Excess moisture is released into the air, which will contribute to the high moisture content in the building.

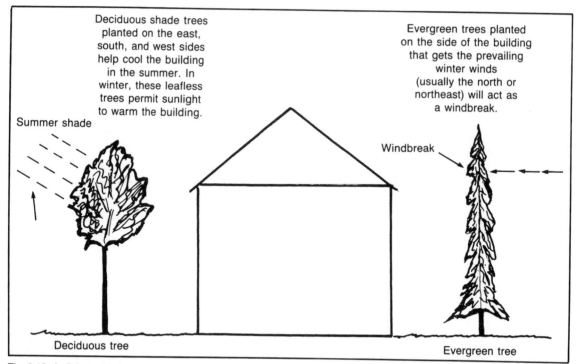

Deciduous shade trees planted on the east, south, and west sides help cool the building in the summer. In winter, these leafless trees permit sunlight to warm the building.

Summer shade

Deciduous tree

Evergreen trees planted on the side of the building that gets the prevailing winter winds (usually the north or northeast) will act as a windbreak.

Windbreak

Evergreen tree

Fig. 6-18. Judicious planting of both evergreen and deciduous trees can prove beneficial for both cooling and heating of a building.

Placing trees and shrubs strategically around your building can help cut your heating and cooling costs. Place deciduous trees (trees that lose their leaves in the winter) on the southern side of your building so that they will provide plenty of shade in the summer but allow sunlight to flow into your building in the winter. Evergreen trees grown on the north and west sides will provide an effective windbreak during the winter storms and help cut your heating costs (Fig. 6-18).

SUMMARY

Nothing written on these pages will prove to be earth-shattering. It is just good common sense. Take your time and look around your building. Are the things that were discussed in this chapter in place and part of your energy defense system? If your answer is no, then get cracking and start your energy maintenance program on its way to success. The hints and suggestions that were offered are only the tip of the iceberg. You can devise many simple savings for your particular property that will make you more comfortable and will also save you money—no matter what season it is. You only need the will to want to do it.

There are many good books on energy conservation on the market—some of which are listed in Appendix B. Search through these books for ideas that can be beneficial to you. Nothing comes easy, of course, and you will have to put in some effort along with putting out some sweat. At least you won't be putting out a lot of money on fancy new gadgets that might not even have any payback.

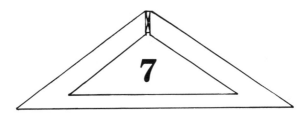

Maintaining the
Mechanical Systems

Because your electrical-mechanical systems work for you every day of each week, maintaining them will be an ongoing proposition. You have to work against the bad tidings of overheating, wear and tear, corrosion, and general deterioration that necessitates a continuous vigil and an organized maintenance program. Without this preventive maintenance, the systems will age prematurely because components will not last their anticipated life expectancies, as seen in Table 7-1.

In your ongoing maintenance program, you must know what to look for during your inspection tours. You also must know what approaches to take to keep your systems in good working order. Read this chapter carefully as well as the details given in Chapter 9.

HEATING AND COOLING SYSTEMS

Let's start with the heating and cooling systems because they are probably the largest and most costly to replace. As you recall, Chapter 6 discussed these systems under energy conservation. Chapter 7 discusses them with respect to maintaining them for long life.

You already know that there are many types of heating systems. Steam, hot water, and warm air are the most common, with some others also in use throughout the country. Electric heat can be found in regions where electricity is inexpensive; heat pumps are used in areas where the temperatures don't drop much below 45° F for too long; and solid fuel stoves are used in regions where wood or coal are relatively inexpensive. This chapter will cover all the important maintenance chores of all the heating systems as well as air conditioning.

**Table 7-1. General Life Expectancies
of Electrical- Mechanical Systems and Components.**

Item	Functional Life Expectancy
Quality cast-iron boiler	30 to 45 years
Average steel boiler	20 to 25 years
Warm-air furnace	25 to 35 years
Combustion chambers	10 to 15 years
Burners	10 to 20 years
Compressors	8 to 10 years
Solid fuel stoves	10 to 15 years
Motors	5 to 7 years
Pumps	5 to 7 years
Heat pumps	15 to 25 years
Air-conditioning systems	15 to 25 years
Galvanized heating and cooling ducts	50 plus years
Humidifier	7 to 10 years
House wiring (copper/aluminum)	Lifetime
Circuit breaker panels	30 to 40 years
Circuit breakers	25 to 30 years
Septic tank leaching field	18 to 22 years
Buried oil tanks	15 to 25 years
Above-ground oil tanks	30 to 40 years
Internal tankless coils	10 to 15 years
External tankless tank	10 to 20 years
Free-standing hot-water tank	5 to 15 years
Sump pump	5 to 7 years

STEAM

A steam system, as illustrated in Fig. 7-1, has several key maintenance areas to check, which will be discussed here in detail. First let's review what a steam system is and what it does.

Steam is generated in a boiler that then rises to the radiators. Once the steam hits the radiators, it forces trapped air out of the radiators through vents, as seen in Fig. 7-1, then gives off its heat. The steam then condenses in the radiators and flows back to the boiler in the form of water, where the entire process begins again.

Maintaining a steam system is easy as long as you know what to do. Unlike a hot-water system, steam will require that you be more alert and perform a few additional maintenance functions during the heating season. The sight glass (Fig. 7-2) is a crucial component because it registers the amount of water in the boiler. The sight glass should normally be three-quarters full of water. If the sight glass is full of water, that means that the boiler has too much water in it and could overflow into the radiators. Too little or no water will cause the boiler to malfunction and possibly crack. So, it is important to periodically check this sight glass to make sure that the right amount is registering. If you find that the water level is low, just open up the fill valve, which is usually located on the water line to the boiler, and fill the sight glass to the correct level. If there is too much water, do the opposite and drain the sight glass by blowing down the drain cock on the boiler.

Another important component of the system is the low-water cutoff valve. This safety device (Fig. 7-3), will shut down the system if the water level gets dangerously low. Inside the valve is a float; when the float reaches a predetermined level, the valve automatically switches off the electrical current to the burner, which in turn shuts down the boiler.

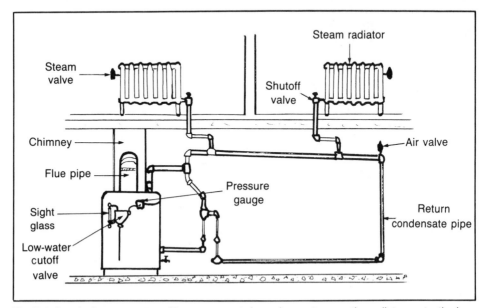

Fig. 7-1. A typical steam system. Key components in such a system are the radiator vents, the low-water cutoff valve, and the sight glass.

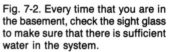

Fig. 7-2. Every time that you are in the basement, check the sight glass to make sure that there is sufficient water in the system.

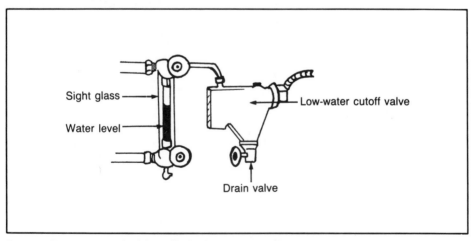

Fig. 7-3. Be sure to drain (blow off) the low-water cutoff valve according to the manufacturer's recommendations. At minimum, test this crucial safety feature once a month.

If built-up sediment and sludge is packed into the valve, however, and the float gets stuck and can't drop down, then the burner will continue to operate and most likely damage the boiler.

Part of your preventive maintenance program, then, is to see to it that the float remains operational. Depending upon the model, purge the low-water valve weekly or semimonthly. Quite often there is a tag on the valve that tells how frequently you have to drain it. If you are not sure, check with your serviceman. All you have to do is open up the drain, as seen in Fig. 7-3, and drain out any built-up sediment into a bucket placed below it. When finished, be sure to bring the water level back to its original setting in the sight glass.

Radiator vents automatically purge air out of the radiators, but sometimes these won't work because of dirt or paint blocking their openings. If the vents aren't working, the radiator won't fill up with steam. Once these vents (Fig. 7-4) become defective, you should replace them. They are usually inexpensive and can easily be installed into the radiator. The shutoff valve on a steam radiator (Fig. 7-5) should either be off all of the way or on all of the way. If not, the radiator might leak at the radiator connections. As part of your maintenance program, therefore, be sure to check before the heating season to make sure that the valves are turned on.

If your heating system is a steam system, review these notes every so often. To recap, periodically inspect the sight glass; blow off the low-water cutoff according to the manufacturer's directions; replace defective radiator vents; make sure that all of the radiator shutoff valves are fully on prior to the heating season, and as is true with all systems, have your serviceman annually inspect and tune up your system.

HOT-WATER HEATING SYSTEMS

There are two types of hot-water heating systems: *gravity hot water* and *forced hot water*. Gravity hot water is found in older homes and is usually not installed today because it is less efficient than forced hot water. A gravity hot-water system circulates

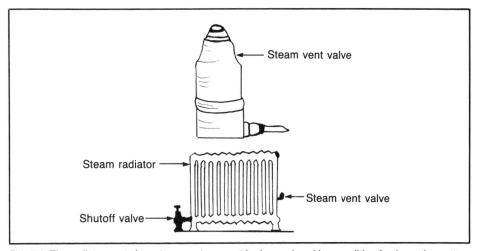

Fig. 7-4. The radiator vents in a steam system must be in good working condition for the entire system to function properly. Replace defective vents.

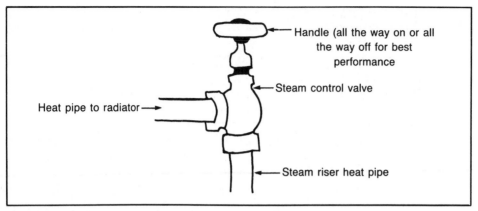

Fig. 7-5. A shutoff valve on a radiator should either be all the way open or all the way closed. The radiator will not operate correctly if the valve is in the midway position.

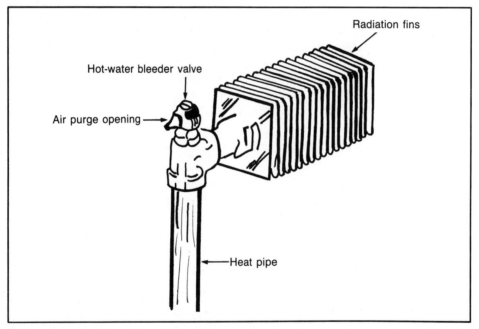

Fig. 7-6. Use a special tool (key) that most hardware stores sell to open the vent valve and bleed a hot-water radiator of trapped air. Once the air is out, continue to drain a half cup of water from the radiator to ensure that all of the trapped air is out.

the heated water in the boiler to the radiators by gravity. As water in the boiler is heated, it becomes lighter in weight and rises through the heat pipes to the radiators, where it gives off its heat. When the hot water hits the radiators it cools down, becomes heavier once more, and flows back down to the boiler to begin the cycle all over. This is an extremely uncomplicated system, one that requires no special wiring other than for the burner and the controls. It is therefore rather simple to maintain. The only thing you

Fig. 7-7. A typical circulating pump on a forced hot-water system. Be sure to oil each of the oil cups with a few drops of light oil (#20) twice a year.

should do is to periodically vent the radiators of all trapped air, as seen in Fig. 7-6.

In a forced hot-water system, a circulating pump (Fig. 7-7) forces the heated water from the boiler through the heat pipes up to the radiators. This system has two advantages over the older gravity system: it is capable of circulating heat to any part of your building and it is more efficient to run. The major maintenance chore that you should be concerned with is to periodically (semiannually) oil the circulating pump's oil cups, as seen in Fig. 7-7. Use a light (#20) oil—a few drops in each cup will do the job. If you hear odd sounds coming from the pump, and the oiling hasn't helped, have your serviceman check it out. It might be replacement time for a pump when it starts giving off signals of distress.

A relief valve on a hot-water system is there to ease the water pressure when it exceeds the recommended high limit, usually 30 pounds. During your maintenance tours check to see if the drip pipe on the relief valve is dripping water. If it is, then check the pressure gauge on the boiler (Fig. 7-8). Usually a pressure reading of 30 pounds or over will cause this problem. Have your serviceman check it out. You either need the relief valve replaced or the expansion tank (Fig. 7-9) bled. These two chores are best left to your serviceman.

Maintaining your hot-water system (either gravity or forced) is relatively simple. Periodically bleed the air vents on the radiators to release trapped air as needed throughout the heating season. Lubricating circulating pumps and monitoring relief valves and pressure gauges are added chores on a forced system. Again, as noted with the steam system, have a professional annual tune-up.

Fig. 7-8. A pressure gauge on a hot-water system. Monitor the gauge during the heating season to make sure that the pressure does not start pushing towards the 30-pound pressure zone. Note the word *danger* adjacent to the 30-pound reading.

WARM-AIR HEATING SYSTEMS

Like hot-water systems, warm-air heating can be divided into those of the older gravity type and the more modern forced warm-air systems (Fig. 7-10). The first works pretty much on the same principle as does the gravity hot-water system, except the source of heat is air rather than water and the means of transporting it are ducts rather than pipes. As is true with the forced hot-water system, a forced warm-air system is again more efficient to operate because the air is circulated more effectively with a blower system. If you still depend upon gravity, perhaps you should check with your serviceman about updating to a forced warm-air system. It is often a cost-effective conversion and one that can be done without having to pull out your old system.

Maintenance on warm-air systems is minimal. Easy steps like keeping your ducts and registers clean and free of lint and dust, and replacing your filters as they become dirty are what it's all about. Annual tune-ups of burners, lubrication of motors and blowers on forced systems, and making sure that your duct system has no openings is usually what it takes to keep your system running effectively.

Part of your ongoing maintenance program is to plug up energy losses as discussed in Chapter 6. Figure 7-11 shows a good place to start for saving money. If you find that the heat ducts have openings through which warm air is being lost, or if you find that the duct tape on the joints has deteriorated, be sure to mark these down for future repairs.

Fig. 7-9. Expansion tanks in forced hot-water systems periodically need to be bled. The shutoff valve seen at the bottom of the tank is where the tank is drained.

A roll of duct tape will pay for itself ten times over when applied to gaps in your heating ducts.

One word of caution about warm-air heating systems that use a humidifier as part of the system (Fig. 7-12). If your system has a humidifier in it, be prepared to have problems with rust in your furnace, particularly in the heat exchanger, which could result in the sad fact that you might have to replace the entire furnace due to corrosive damage. If you already have a humidifier in your furnace, have your serviceman check it out and advise you on any corrosion that might be building up. If you feel that you need additional humidity in your building, consider individual room humidifiers instead of one in your furnace.

Once every six months, check the condition of the fan belt on the blower (Fig. 7-13) to see if it is too loose. Press your thumb against the belt; if it deflects more than ½ inch, have your serviceman adjust the tension. A loose belt or one that is too tight will make the blower-fan work too hard, which could cause the unit to overheat. Also check for deterioration in the belt; if extensive damage is noted, replace the fan belt.

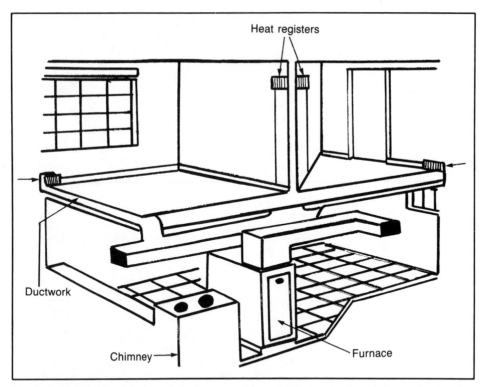

Heat registers

Ductwork

Chimney →

Furnace

Fig. 7-10. A typical forced warm-air (hot) system. It is probably the simplest of the modern heating systems and requires minimal maintenance and service.

Maintaining warm-air systems can be summed up in a few words. Keep your filters, registers, and ducts clean and free of dust, lint, and dirt. Seal up open joints in ductwork with duct tape to avoid heat losses. Replace filters and fan belts when their time is up. Avoid using humidifiers in warm-air systems, and if you already have one, be sure to have it professionally inspected for corrosive damage to the heat exchanger. And don't forget to have annual tune-ups by your service company.

MAINTAINING OIL BURNERS

You do want your oil-fired system (steam, hot water, warm air) to run efficiently, so don't try to save money by not calling in a licensed technician. Rather, have a tune-up every year. More often than not your fuel company will remind you that it's time for one. The tune-up involves making fine adjustments to the burner nozzle assembly and to gauge the airflow to the combustion chamber. Efficiency testing during and after the tune-up ensures that the system is working to the best of its capacity. Figure 7-14 shows a typical testing instrument for calibrating heating efficiency.

While the oil company serviceman is there, ask about the value of reducing the firing rate of the burner. Most burners are oversized and therefore oil guzzlers. Minor adjustments, such as replacing the nozzle with a smaller size, can substantially increase the efficiency of the burner and save you a lot of money. Of course, some oil companies don't mind oil guzzlers, so they might not volunteer to make such changes. Also make

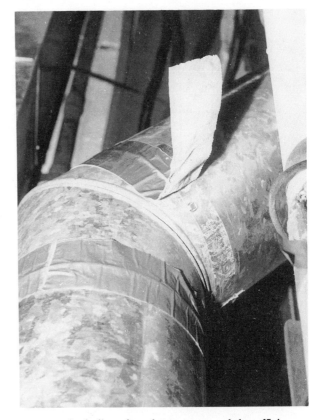

Fig. 7-11. Be sure that all heat duct joints are properly sealed with duct tape to avoid heat losses in basement or crawl-space areas.

sure that the serviceman leaves a tag on the boiler after the tune-up and the efficiency test. The tag should show the final combustion efficiency rating.

Here are the usual areas included in an oil burner tune-up. Depending upon the type of system that you have, some of the areas listed might not apply. Make your serviceman explain what he is doing and why.

- Combustion chamber inspected, repaired, or replaced
- Heat exchanger inspected and cleaned
- Oil pump inspected and cleaned as well as regulated
- Oil filter cleaned or replaced
- All operating and safety controls operated and checked
- Pumps, fans, blowers, and motors checked and lubricated
- Draft regulator checked and replaced if necessary
- Burner components inspected, cleaned, and lubricated
- Oil pump bled if required
- Nozzle replaced annually
- Heating system inspected for air or water leaks
- Oil tank inspected for leaks and corrosion
- Flue pipes checked for corrosion and cleaned
- Chimney inspected for deterioration and cleaned if needed
- Efficiency tests performed during and at completion of tune-up

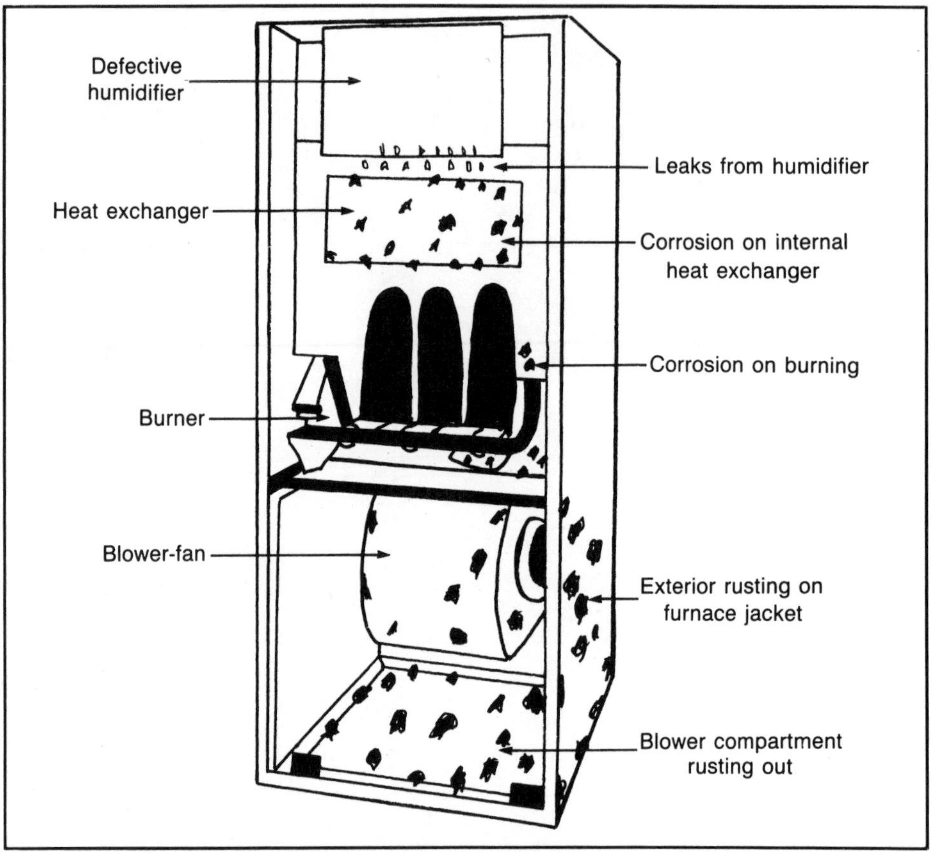
Defective humidifier
Leaks from humidifier
Heat exchanger
Corrosion on internal heat exchanger
Corrosion on burning
Burner
Blower-fan
Exterior rusting on furnace jacket
Blower compartment rusting out

Fig. 7-12. Humidifiers installed in warm-air systems have been known to cause extensive damage and corrosion in the system and sometimes result in the costly replacement of the furnace.

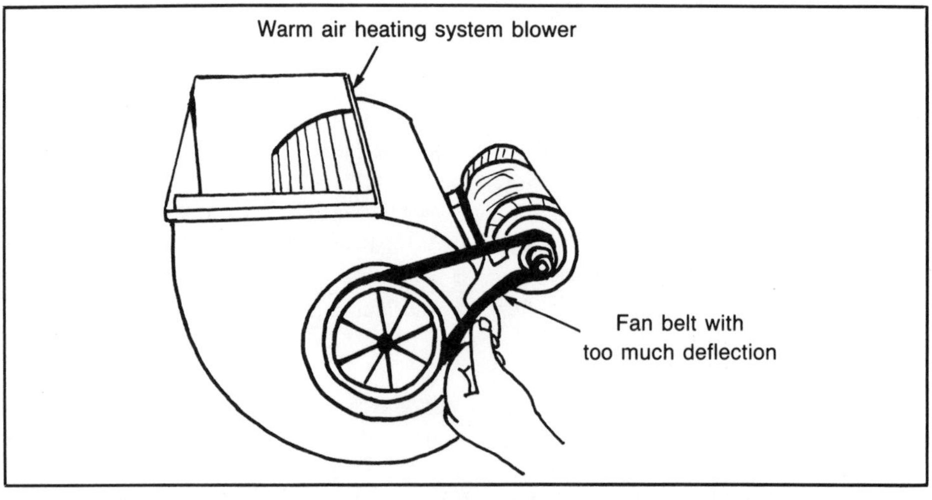
Warm air heating system blower
Fan belt with too much deflection

Fig. 7-13. Once a year test the tension on the fan belt by pressing in on it with a finger. If you have more than ½-inch deflection, have your serviceman adjust and tighten it up.

Fig. 7-14. Once a year have your serviceman perform an energy efficiency test on your system. Testing equipment can save you money in the long run.

MAINTAINING GAS BURNERS

Gas burners, unlike oil burners, do not need annual tune-ups unless the equipment warrants it. To ensure its efficiency, though, have your system tuned up every other year and have it done either by your gas company or by a licensed heating contractor.

The two main areas of the gas tune-up include basic cleaning and adjusting the air and gas mixture on the burner. All scale, rust, and dirt must be removed from the heat exchanger, the gas jets in the burner, and from the air inlets on the burner. The air and gas mixture for the burner must be finely adjusted to give the optimum combustion with the minimum carbon monoxide. As with the oil tune-up, efficiency testing should be done throughout the tune-up period and a final efficiency test should specify the exact efficiency rate of your system.

Listed below are the areas that are included in a gas burner tune-up. Be sure to ask your serviceman any questions that you feel you need answers to, particularly if you have concerns about safety.

- Heat exchanger cleaned and inspected
- Flue pipes checked for corrosion and cleaned

- Air inlets cleaned and adjusted
- All operating controls and safety features operated and checked
- Venting ducts, gas lines, gas fittings, and gas valves inspected for leaks, corrosion, and deterioration
- Pumps, fans, blowers, and motors checked and lubricated
- Pilot safety valve, automatic gas valve, and pressure regulator inspected and serviced

One last point about gas burners. During the energy crunch a few years back, many new techniques came into vogue on saving energy, one of which was to turn off the gas pilot light on the gas burner during the off-heating season. Please don't make that error. Although you might save a few dollars by putting out the gas pilot light, studies have shown that if your basement has a high concentration of humidity and moisture, you will be losing money in the long run for repairs to your furnace or boiler because of corrosion.

ELECTRIC HEAT

Because electric heat is clean and relatively trouble-free, little has to be done to keep it functioning efficiently. Here your major concern is insulation in the building. Electricity is relatively expensive to heat with, no matter where you live, so you want to make sure that you have a tight building. Even in areas where electricity is cheaper than oil or gas, you should consider having at least 12 inches of insulation on attic floors and a minimum of 3½ inches in walls—preferably 6 inches. It is also a good idea to have a back-up heating system, such as a wood stove, just in case you lose power during a cold spell.

SPACE HEATERS

Make sure that it is okay to use gas or kerosene space heaters in your building before you heat with them. Your local codes should specify which units are allowed to be used in buildings and which cannot be used. If you are using a space heater that is not approved, stop heating with it because you might be taking your life into your own hands. Chapter 8 lists tips on how to avoid such hazards.

Because proper venting of such appliances is a must, make sure that the flue pipes are secure and that they properly vent combustion gases to the outside. If you have any questions or concerns, call your local building department. Here are some simple safety tips for space heaters. You will find additional ones in Chapter 8.

- Make sure that all rooms with space heaters are well ventilated.
- Place heaters at least 3 feet from doorways, hallways, and combustible materials.
- Use only UL-approved heaters.
- Never permit children to operate or refuel heaters.
- Vent gas space heaters to approved flues.
- Never leave a kerosene heater unattended.
- Never place space heaters in bedrooms.

SOLID FUEL STOVES

If you already own one or if you are planning to buy a solid fuel stove, you must read the very fine publication published by the National Solid Fuel Council. This booklet illustrates all of the code specifications for free-standing and insert-type stoves. There isn't a question that you could possibly have that isn't answered in this manual. For information on where to get it, refer to the Appendix.

To make sure that there is no way that your stove can cause a fire, you should have no combustible materials near the stove or the flue pipe—absolutely none. Be sure to check very carefully during your inspection-maintenance tours. Figure 7-15 illustrates the common clearances for most wood-burning stoves.

If a solid fuel stove is the only system or the main source of your heat, it is particularly important that you have a regular schedule of cleaning the stove, the flue pipe, and the chimney flue liner. Cleaning every month or two months is not outrageous (depending upon the usage of the stove) but highly recommended. It's best to leave important maintenance to a professional, especially if you have a high roof to climb to get to the chimney.

Use the helpful fuel guide for wood-burning stoves in Table 7-2 to help you make good selections for wood fuel. Remember, never burn chemically treated wood because noxious odors can escape through faulty seals in the stove or flue pipe system and enter the building. Try to burn only dry, well-seasoned firewood. not only will this give you a better fire but also will eliminate major creosote buildup. Remember that solid fuel stoves are potentially the most dangerous of heating systems. A detailed analysis can be found in *What's It Worth? A Home Inspection and Appraisal Manual* (TAB Book #1761).

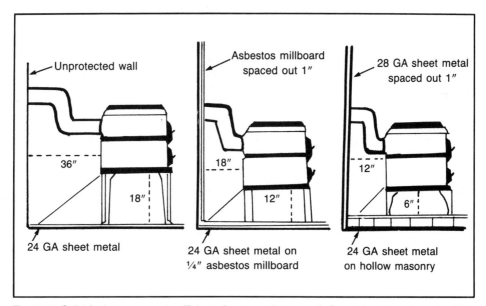

Fig. 7-15. Solid fuel stoves need sufficient clearances between their metal parts and combustible materials. Use this illustration as a guideline with your system.

Table 7-2. Fuel Guide for Solid Fuel Stoves.

Fuel	Characteristics
Hickory	High heat value. Burns completely.
Oak	High heat value. Hard to split. Needs to be well seasoned.
Maple	High heat value. Burns fast.
Beech	High heat value. Colorful flame.
White ash	Moderate heat value. Splits easily.
White elm	Moderate heat value. Burns poorly unless well seasoned. Difficult to split.
White birch	Moderate heat value. Burns completely.
Douglas fir	Fair heat value. Very smoky fire.
Chestnut	Fair heat value. Smoky fire with frequent sparks.
Spruce	Fair heat value. Burns fast; smoky and sparks.
Cedar	Fair heat value. Pleasant aroma; smoky and sparks.
White pine	Poor heat value. Burns fast without sparks; smoky fire.
Sawdust logs	Fair heat value. Colorful flames; relatively expensive.
Newspaper logs	Poor heat value. Smoky fires.
Anthracite (hard coal)	Moderate heat value. Difficult to ignite. Burns slowly; sootless and smokeless.
Bituminous (soft coal)	Fair heat value. Burns fast; sooty and smoky.
Cannel coal (soft coal)	Poor heat value. Ignites quickly; sputters.

HEAT PUMPS

Heat pumps (Fig. 7-16) are devices that work like air conditioners, only in the reverse. During the summer, they take heat out of your building and dump it outdoors; during the winter, the reverse is true—they pick up heat from the outdoors and bring it in to warm your building. They have one drawback, however, and that is the further north you live, the less effective they are. Oftentimes you need a backup system,

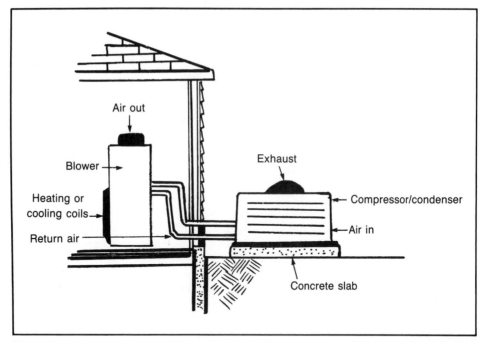

Fig. 7-16. A heat pump system might not be suitable for your region of the country. States that are too far north might not have sufficient warm winter days to allow maximum efficiency from these novel heaters.

particularly in regions of the country where the temperatures drop below 45° F. Electric heat as well as solid fuel stoves are often used to complement this novel way of heating your building.

Because a heat pump works much like an air-conditioning system, your maintenance chores will be similar. In addition to following the maintenance tips given in this chapter, be sure to always follow the manufacturer's recommendations for service and upkeep. Chapter 9 will have some recap maintenance hints as well.

AIR CONDITIONING

Probably the most important thing to remember here is to have your system serviced and checked out every year by a reputable service company or by the dealer that installed your system. That's not the only thing you have to do, though. Before the cooling season begins, replace filters as needed. Also remove debris from around the compressor and trim back shrubbery that might provide too much shade. Look over the compressor slab (Fig. 7-17) for signs of settlement or cracks. Check insulation on pipes for deterioration and, if necessary, make repairs. Caulk and seal all holes that were made in the building walls for the piping to keep out the weather and insects.

When your central air-conditioning system is serviced, make sure that the serviceman does the following:

- Cleans and/or replaces air filters (you can do this during the cooling season yourself)
- Checks and vacuums the evaporator and condenser coils

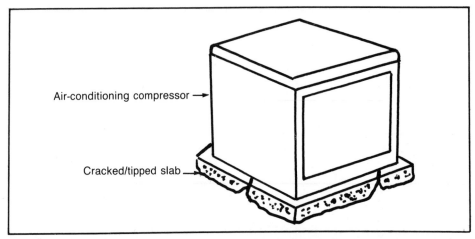

Fig. 7-17. Cracks and displacement in a concrete slab for an air-conditioning compressor or a heat pump could cause some tipping of the units and damage to connected piping.

- Inspects, cleans, and lubricates the evaporator and condenser motors
- Inspects and cleans the fans and checks the fan belts to see if they need to be replaced.
- Checks air grilles for obstructions (you can do)

See Table 7-3 for an Air-Conditioning and Heat Pump Troubleshooting Chart. Refer to Table 7-4 for a recap of heating maintenance.

ELECTRICITY

Is the electricity in your building a good servant to you or a dangerous master? The only way you can tell is if you periodically test all the components that make up your electrical service. Before you can start testing, though, you might have to buy some inexpensive electrical testers. The combination outlet analyzer on the right in Fig. 7-18 and the continuity test on the left make it possible for you to check all of the outlets in your building for most types of defects. Table 7-5 lists some of the defects that your testers can check, such as ungrounded outlets and reversed polarity. Each of these testing devices comes with simple directions. As is true with any defect in your electrical system, if you don't know how to repair it, by all means hire a licensed electrician to do it.

Just because your lights go on when you flip a switch doesn't mean that your system is in good condition. On the contrary, loose wires, corrosion on wire connections, overheating wires, overfusing at the service panel, and a host of other defects could exist in a seemingly perfect electrical system. With this in mind, be sure to carry on with an ongoing inspection-maintenance program, especially with your electrical service.

Once a year, pull or trip your main disconnect switch to make sure that corrosion has not built up to the point where it won't turn off the power during an emergency. Every six months, trip all the circuit breakers in the service panels. Circuit breakers need to be "exercised" periodically, and you can do that by switching them on and off several times. This will keep them in working order.

Table 7-3. Chart for Air-Conditioning and Heat Pump Troubleshooting.

Problem	Cause	What To Do
System does not operate	Faulty thermostat	Replace
System operates continuously	Defective switch	Replace
	Thermostat in wrong location	Reposition
System goes on and off repeatedly	Thermostat in wrong position	Reposition
System does not cool air adequately	Dirty filter	Replace or wash
	Clogged registers	Clean
	Faulty thermostat	Replace
	Blower belt loose	Tighten belt
Blower motor overheats	Needs lubrication	Oil motor oil cups
	Blower belt too tight	Loosen belt
Excessive noise or vibration	Incorrect belt tension	Adjust tension
	Blower motor loose	Tighten mounting bolts
	Loose grilles or access panels	Tighten screws or secure with duct tape

Remember this is only a brief review of some typical problems. Be sure to have a professional service and maintain your systems.

If your building is equipped with ground fault circuit interrupter outlets (Fig. 7-19), test them on a monthly basis or as often as recommended by the manufacturer. Do this by tripping the test button and then resetting it. If after testing you find that it does not trip out, have an electrician replace it. If you presently do not have these lifesaving safety devices in your building, consider having an electrician install them. Bathrooms, the kitchen, basement, garage, outdoor receptacles, and if you have one, a swimming pool, are all places that these devices should be installed.

Quite often you might find yourself short in the number of outlets that you need, so you might decide to extend them by using lamp cord wiring. Please don't do this. This arrangement is potentially very dangerous, at best (Fig. 7-20). This type of electrical hook-up is a leading cause of fires and electrical shocks. If something like this condition exists in a building that you own, have an electrician correct it by providing ample outlets for each room.

Also, and this is very important, don't leave live wires exposed, as seen in Fig. 7-21. Enclose all live wires in proper receptacles or junction boxes to prevent anyone from accidentally touching them. In addition, cover junction boxes to again provide the same safety feature—to avoid electrical contact.

Table 7-4. Maintaining Heating Systems.

✔ Monitor steam system sight glass on weekly basis.
✔ Blow off low-water cutoff valve on weekly basis.
✔ Refill steam boiler when sight glass is low.
✔ Replace radiator valves or vents when required.
✔ Keep radiator shutoffs either all the way on or off.
✔ Bleed hot-water radiators as required (beginning of heating season).
✔ Lubricate oil cups on motors and circulating pumps.
✔ Monitor relief valves and pressure gauges during the heating season.
✔ Have expansion tanks recharged by a serviceman when required.
✔ Replace or clean filters for warm-air systems and heat pumps.
✔ Oil and lubricate motors and blowers on warm-air systems.
✔ Vacuum registers and ducts on warm-air systems.
✔ Avoid using a humidifier in a warm-air system.
✔ Monitor fan belts for deterioration and looseness.
✔ Have burners professionally cleaned and serviced on annual basis.
✔ Have heating system inspected and serviced on annual basis.
✔ Remove debris and vegetation from heat pump compressor.
✔ Check and repair insulation on heat pipes and ducts.
✔ Follow the manufacturer's recommendations for service.
✔ For electric heat, make sure there is sufficient insulation in building.
✔ With space heaters, be sure to follow local code requirements.
✔ Clean and inspect solid fuel stoves on a regular basis.
✔ Always follow local safety codes and regulations as well as common sense when dealing with any heating appliance.

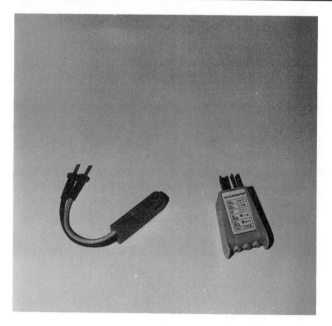

Fig. 7-18. Buy yourself some inexpensive electrical testers and use them to check electrical outlets for wiring defects.

Table 7-5. Possible Outlet Defects.

✓ Outlet has no power.
✓ Outlets are loose in walls.
✓ Outlets are missing covers.
✓ Outlet covers are warm or hot to the touch.
✓ The covers on outlets are metal rather than plastic.
✓ Outlets are ungrounded.
✓ Outlets have a hot-neutral reversed polarity.
✓ Outlets have a ground-neutral reversed polarity.
✓ Outlets have a hot-ground reversed polarity.
✓ Ground fault outlets might be defective.

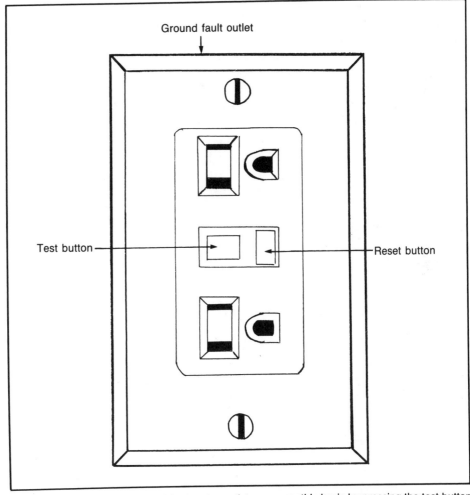

Fig. 7-19. Test ground fault circuit interrupter outlets on a monthly basis by pressing the test button and then resetting the outlet.

Fig. 7-20. Avoid using makeshift wiring such as lamp cord outlets. If you need additional circuits, have an electrician install safe wiring and outlets to meet your demands.

Fig. 7-21. Never allow live wires to be in such a position as seen in this photograph. Electrical shock and fire could result from such defective wiring.

Once a year, go to the electric grounding connection (Fig. 7-22). Check to see if the connection to the water pipe is nice and tight. Sometimes a workman might remove it to work on a waterline or a water meter and then forget to replace it. If it is off or loose, be sure to restore it to the water pipe. If you see rust buildup on the clamp connection, be sure to wire brush it off as this corrosion could interfere with the grounding connection.

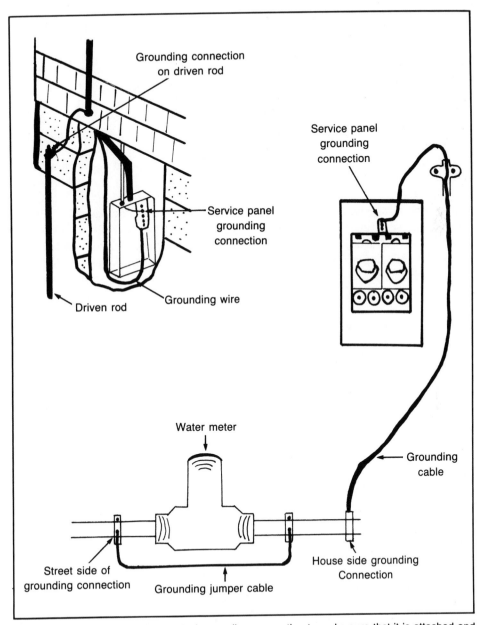

Fig. 7-22. Annually check the electrical grounding connection to make sure that it is attached and firm on the water pipe or driven rod.

In your service panel, always check to make sure no one has installed a higher rated fuse to a fuse socket—for example, putting in a 30-amp fuse in a fuse socket that only should have a 15- or 20-amp fuse. This overloading of a circuit could cause a fire, so be sure to reduce the fuse size. Also, never install a copper penny or a copper tube in place of a fuse. Anyone who does this is asking for an electrical fire. Use common sense when dealing with electricity!

Loose electrical receptacles and light fixtures are another potential hazard area. Be sure to have any such defects corrected. If you find light switches that work from a metal pull chain, be sure to add a piece of cloth to the end of the chain. An unsuspecting user could be on the receiving end of an electrical jolt. It would even be a better idea to have an electrician change over to a wall switch to control such lights.

Consider modernizing metallic outlet covers and switch covers to nonconductive plastic covers. Metallic covers can and have been known to transfer electric current from a faulty wired outlet so that plugging in your iron, for example, could become a shocking experience.

Never put a light bulb into a fixture that calls for a lower wattage than what you have in your hand. If you fail to heed the safety wattage ratings, you could have a fire on your hands. Change fluorescent bulbs as soon as they start blinking, as this usually is a warning of overheated ballast. If you happen to smell a strong odor of tar coming from a fluorescent light fixture, it could be because the ballast is ready to boil over. In this case, do not use this light fixture until after it has been repaired or replaced.

If you are going to replace an outlet, be sure to remember and use the *BB rule*. Simply, this rule states that you always attach the black wire to the brass terminal and the white wire to the light or silver terminal in the outlet and the grounding wire to the metal housing in the junction box. All of this is illustrated in Fig. 7-23. Remember, *black to brass* when wiring an outlet, and you won't go wrong.

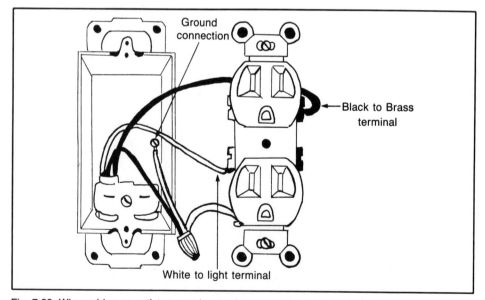

Fig. 7-23. When wiring an outlet, remember to always install the black wire to the brass terminal in the outlet box and the white wire to the light-colored terminal in the box. Remember BB, Black to Brass.

Table 7-6. Rules for Electrical Safety.

✔ Always turn off the power when inspecting or working on your electrical service.

✔ Always test all wiring before you touch it. Use a simple voltage tester that can be bought in most supply stores.

✔ Always wear rubber-soled shoes or sneakers when working with electricity.

✔ Never stand on a wet or damp floor when working with electricity.

✔ Only work during the daylight hours.

✔ Try not to work alone.

✔ Don't work when you are tired and never rush your work.

✔ Correct any safety hazards before working on the system.

✔ Keep a powerful flashlight handy for your inspections and work.

✔ Double-check to see if your tester is working by using it on a circuit that you know is functional.

✔ When replacing parts, only use equipment approved by the Underwriters Labs (UL).

✔ Always assume exposed wiring is live.

✔ Never use lamp cord wiring in place of normal circuit wiring.

✔ Always use ground fault interrupter circuits in hazardous locations.

✔ Only use one hand to insert or remove fuses. Keep the other hand in your pocket.

✔ Never put your finger or a tool into a fuse socket.

✔ Always use a fuse puller to remove cartridge fuses.

✔ To wire outlets, always remember the BB rule (Black to Brass): Black wire goes to the brass terminal, white wire goes to the light terminal.

Use Tables 7-6 and 7-7 for a recap of electrical safety rules and electrical maintenance tips. And remember that the better part of valor is to call in a professional whenever the question of safety arises.

PLUMBING MAINTENANCE

Plumbing maintenance not only includes keeping an eye on water pipes and drain pipes, but also on maintaining septic and cesspool systems, servicing hot-water tanks, inspecting sump pumps, and preventing freeze-ups in water lines.

Septic/Cesspool Systems

Septic and cesspool systems are private sewage disposal facilities used by people that live in areas not serviced by municipal systems. As you can see in Fig. 7-24, a *septic system* is made up of a septic tank, a distribution box, and a leaching field. Household wastes travel from the house sewer line to the septic tank where decomposition begins. Solid wastes settle to the bottom of the tank with the remaining liquid (effluent) staying at the top. When the tank is full, the liquid flows out of the tank and into the leaching

Table 7-7. Maintaining the Electrical System.

✓ Buy a combination outlet and ground fault outlet tester and a continuity tester and use them to test outlets.

✓ Test the main disconnect by tripping or pulling it annually.

✓ Trip all circuit breakers every six months.

✓ Test ground fault outlets on a monthly basis.

✓ Have an electrician repair or replace faulty outlets.

✓ Check the grounding connection for the service panel on an annual basis.

✓ Have ground fault circuit interrupters installed (if you don't have them) in your bathrooms, kitchen, basement, garage, swimming pool, and exterior outlets.

✓ Replace lamp cord wiring with appropriate circuits and receptacles.

✓ Have an electrician install additional circuits and outlets in rooms with minimal service.

✓ Provide covers for junction boxes with exposed wiring.

✓ Install all exposed wiring in appropriate junction boxes.

✓ Reduce fuse sizes in overfused circuits.

✓ Secure loose ceiling lights and wall receptacles.

✓ Replace metal pull-chains with cloth pulls or change to switches.

✓ Substitute plastic outlet covers for metal ones.

✓ Replace light bulbs according to their wattage ratings.

✓ Change fluorescent bulbs when they start blinking.

✓ Caulk and seal exterior electric meter connections to building.

✓ Provide waterproof covers for outdoor receptacles.

✓ Provide safety covers on interior outlets to childproof them.

✓ *Follow the electrical safety guidelines in Table 7-6.*

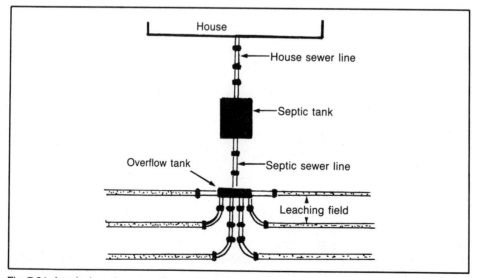

Fig. 7-24. A typical septic system. Remember to have your system inspected once a year and pumped out every 3 to 5 years.

field by way of the distribution box. Here it seeps out through perforated pipes into the adjacent soil.

Cesspools, on the other hand, are simply pits with earth bottoms and sidewalls constructed of stone or hollow concrete blocks. Sewage flows directly to the cesspool with the heavier solids settling to the bottom and the lighter ones, such as grease, floating near the top. Liquids seep through the openings of the sidewalls into the surrounding soil. Cesspools have a limited life span, and a failing cesspool should be replaced with a proper septic system.

Maintaining private sewage disposal systems (cesspool and septic systems) requires some basic maintenance, as shown in Table 7-8. Inspect tanks on a yearly basis to see if they are overloaded with waste products, and if so, have them pumped out. Under normal circumstances, tanks should be pumped out every 3 years. If you have an undersized tank for your building size, annual pumping might be necessary. Usually a 1,000-gallon tank can handle a 3-bedroom home.

As part of your ongoing maintenance of septic or cesspool systems, make sure that no chemicals, cleaning agents, acids, pesticides, paints, and disinfectants, particularly antiseptic agents like lye, are flushed down drains or toilets. These substances can kill or deactivate the microorganisms that help make your system work. Make sure that everyone who uses your building understands this.

If you have a garbage disposal tied into the system, use it sparingly. Better still, have it removed to avoid overloading your tank and leaching field with excess fats and grease.

Don't flush things down your drains that will be difficult to degrade: disposable diapers, facial tissues, colored toilet paper, cigarette butts, coffee grounds, cat litter, sanitary napkins, and plastic items. It's best to throw such things out with the trash.

Avoid pouring cooking oils, fats, and grease down kitchen drains. These substances accelerate the clogging of the leaching fields and should be disposed of with your garbage.

Provide dry wells to take care of the water discharge from dishwashers and washing machines. Too much water in your system will push sewage out of the tank before it has had sufficient time to be digested by the microorganisms working there. You should be especially stingy with water during times of the year when heavy rains or melting snows saturate the ground, reducing the soil's capacity to absorb water.

Table 7-8. Septic System Maintenance Tips.

✔ Inspect systems on a yearly basis.
✔ Have tank pumped out every three years, or sooner if required.
✔ Don't pour harsh chemicals down drains.
✔ Limit or avoid using garbage disposals.
✔ Avoid flushing nonbiodegradable materials into system.
✔ Don't drain cooking oils, fats, or grease into your system.
✔ Provide dry wells for washing machines and dishwashers.
✔ Reduce unnecessary use of toilets.
✔ Be conservative with laundry detergents and bleach.
✔ Don't park vehicles over the tank or the leaching field.

Try to reduce unnecessary toilet discharges to the septic or cesspool tanks. Some people have a bad habit of throwing something like a cigarette butt into a toilet and then flushing it away. This thoughtless action not only helps to overload a system, but also wastes precious water. When you consider that some of the older toilets use up to 6 gallons of water per flush, you might want to consider replacing it with a newer water-conserving one.

Be conservative with the amount of laundry detergents and bleach that you use and try to wash with liquid soaps rather than with the powdered kind because liquid degrades more readily. Also, if your system (tank-leaching field) is saturated with water, try to spread your laundry work out over the week instead of doing it all in one day.

Never drive or park cars and trucks over the tank or the leaching field. Their weight will compact the soil, reduce the soil's ability to absorb moisture, and damage the drain pipes or tank.

Once you get these basics down, you will realize that there is no magic about keeping a private sewage disposal system in good working order. Review the maintenance tips in Table 7-8 as a reminder. A particularly useful booklet, *Everything You Wanted To Know About Septic Tanks But Were Afraid To Ask*, is available by writing to Church & Dwight Co., Inc., 20 Kingsbridge Road, Piscataway, NJ 08854, or by calling (201) 885-1220.

Domestic Hot Water

In most homes, the domestic hot water is supplied by a tankless system working off the boiler or by a free-standing tank fueled by gas, electricity, or oil. Each of these has its own peculiarities and must therefore be inspected carefully and maintained accordingly.

If your hot water is supplied by an internal tankless as seen in Fig. 7-25 or by an external tankless as viewed in Fig. 7-26, you probably should think about installing a sep-

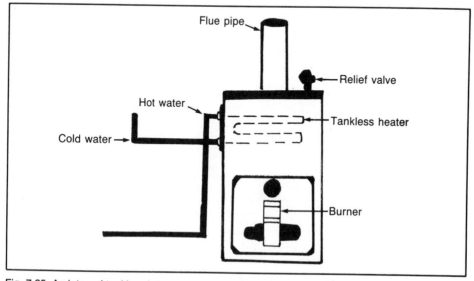

Fig. 7-25. An internal tankless hot-water system. Tankless units are less efficient than free-standing tanks.

172

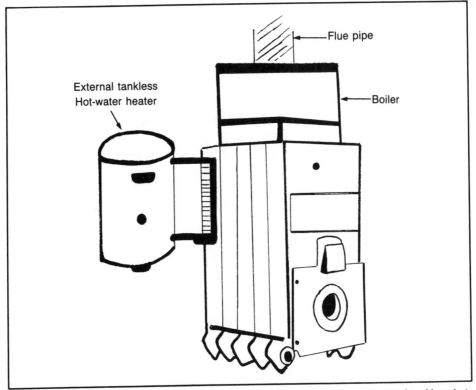

Fig. 7-26. An external tankless unit off a boiler. As with the internal system, external tankless hot-water systems are not as efficient or economical as a free-standing tank.

arate free-standing tank. This will provide you with hot water during the off-heating season without having to activate your heating system. Each of these two illustrated tankless systems often do not deliver an adequate flow of hot water to meet most peoples' demands. (If this applies to you, read on to see what you can do about it.) If, on the other hand, you are satisfied with the internal tankless system that you have, then the only maintenance chore for you to do is to have your service company periodically (every 3 to 5 years, depending upon the water) clean the coils in the boiler or just monitor the external tank for signs of corrosion and leaks.

Even if you are satisfied with your present tankless system, you will want to know that these types are highly inefficient and uneconomical to use during the off-heating season. Efficiency tests have revealed that they run approximately 18 to 22 percent efficient during the summer months. This fact alone might prompt you to add a backup system to use during the off-heating season.

Table 7-9 lists the most common types of domestic hot-water tanks and their characteristics. Part of your inspection-maintenance program is to always check to see if your unit needs to be replaced and to pretty much know in advance what you want to replace it with. Read the information contained in Table 7-9 and refer to it if and when it comes time to make a replacement.

You will be happy to learn that electric hot-water tanks are virtually maintenance free and that they provide long years of use. There is one drawback, however, and that

Table 7-9. Domestic Hot-Water Systems.

Type	Characteristics
Internal tankless found in boilers	Inefficient in supplying hot water when demand is high Uneconomical to run during the off-heating season. Requires periodic cleaning of the internal coils by a serviceman. Tends to corrode and leak after several years of service.
Internal tankless found in some furnaces	Inefficient and uneconomical. Not recommended.
External tankless found on boilers	Inefficient and uneconomical to use during the off-heating season and has limited storage capacity.
Free-standing gas fired tank	Relatively inexpensive to install and operate. Has to be vented to a chimney. 50-gallon tank recommended for a family of four.
Free-standing oil-fired tank	Somewhat expensive to install. Fuel costs will vary from region to region. Requires a chimney to vent to. Requires annual service and cleaning.
Free-standing electric tank	Moderately priced to install. Usually costly to operate unless you are in a region with inexpensive electricity. Clean burning and does not require a chimney to vent to. Minimum 80-gallon tank for a family of four.
Booster storage tank off boiler	Relatively expensive to install and less economical than a free-standing tank. More efficient than tankless. Still requires a boiler to run during the off-heating season.

is that if you live in a region of the country that has expensive electric rates, it could be rather costly to run. You can cut the bills somewhat, though, by insulating your tank and its water lines.

Oil-fired, hot-water tanks require annual cleanings and service of the burner. If you don't have them professionally tuned up annually, not only will you experience more costly breakdowns of components, but the tank will cost you more to operate. So be sure to have your oil serviceman come in once a year.

A gas-fired, hot-water tank is relatively clean in its operation and requires only a minimum of your time. Periodically check around the flue to see if there is evidence of soot or carbon deposits, which is usually the first sign that a good flue or chimney cleaning is warranted. As with any free-standing hot-water tank, drain off a gallon of water every 6 months to avoid sediment buildup in the tank. Figure 7-27 shows you where the drain-off valve is located.

Figure 7-28 shows the water temperature control dial. Make sure that you heat your water on the lower settings of 120° F or 130° F for energy conservation. A higher temperature setting is not only more costly but the hot water can scald you. In addition, a higher setting tends to hasten the demise of a tank. So for all of the above reasons, be sure that you reduce your temperature setting if it is too high.

If your tank is fueled either with gas or oil, noxious fumes that are produced by these fuels must be vented to a chimney. Once a year check the flue pipe of the tank to make sure that it does not look like the one in Fig. 7-29. If you see defects such as

Fig. 7-27. Every 6 months, drain off built-up sediment from the base of a free-standing hot-water tank. If you live in an area with a high mineral/sediment content in your water supply, then monthly draining of at least one gallon should be part of your maintenance program.

Fig. 7-28. The control valve for a hot-water tank. The temperature control dial is seen at the bottom of this control box.

Fig. 7-29. Keep all hot-water flue pipes tightly connected and sealed to chimney connections.

these, be sure to either replace the flue pipe or reseal the flue pipe connection to the chimney with a good application of furnace cement.

Temperature-pressure relief valves (Fig. 7-30) should be tested once a year to make sure that they are working properly. Before you do that, though, make sure that no one is standing near the discharge area of the valve. In Fig. 7-31 you see an excellent way to allow excess water to drain off with a drip pipe. If your relief valve does not have such a drip pipe, consider adding one. To test the valve, gently lift up the lever; a slight gush of water should discharge from it. If no water comes out, have a plumber replace the valve.

Should your relief valve drip, be sure to place a container under the drip pipe to catch the water. Mark this defect down on your maintenance schedule and continue to monitor for further leaks. Consider having the valve replaced if the dripping continues for any length of time.

Never store gasoline, spray cans, or materials that emit volatile vapors near water heaters because the heater's pilot light or burner could ignite them and set off a fire. Also don't hide your heater in a closet or confined area, if it can be helped. Hot-water

Fig. 7-30. The temperature relief valve on the right side of the tank. Note the drip pipe connected to it; it will discharge water to the floor and not at anyone standing next to the tank.

Fig. 7-31. Ideally the drip pipe of the relief valve should be discharged where it can drain into the plumbing system.

tanks need ventilation, otherwise they can't work properly and efficiently. Confined areas restrict the flow of air that is necessary for complete combustion. If yours is in a closet, provide louvers in the door for venting the enclosed area.

Finally, don't let your enthusiasm with insulating everything that doesn't move get the best of you. Keep insulation away from the flue pipe, the controls, and relief valve. And if you must insulate over the instruction data plate, be sure that you can easily get to it if you need to.

If you have a gas-fired tank and you smell gas, follow the listed guidelines given below:

- Open windows.
- Don't operate electrical switches.
- Extinguish open flames.
- Call the gas supplier from a neighbor's house immediately.

As you monitor and check your tank from time to time, note rust spots and areas of corrosion. Record on your maintenance schedule any such areas and double-check

Table 7-10. Tips for Hot-Water Tanks.

✓ Insulate hot-water tank and pipes.
✓ Don't cover flue pipes or access panels with insulation.
✓ Service oil-fired tanks on annual basis.
✓ Keep flue pipes and chimney passages free of obstructions.
✓ Reseal flue pipe connection to chimney with furnace cement.
✓ Drain off sediment buildup in tank every three to six months.
✓ Run tanks on lower temperature settings for higher efficiency.
✓ Vent gas and oil hot-water tanks to chimneys.
✓ Test temperature relief valves once a year and replace defective ones.
✓ Have serviceman check valves that constantly leak.
✓ Never store flammable liquids near hot-water tanks.
✓ Don't install heaters in confined areas.
✓ Provide sufficient ventilation for hot-water tanks.
✓ In garages, protect tanks from vehicle damage.
✓ Monitor old tanks for leaks.
✓ Anticipate replacement when the tank reaches its estimated life expectancy.

the age of the tank. Replacement time might be coming up shortly. As you do your maintenance chores, always refer back to Table 7-10 for a recap of what you should be doing and looking for.

Sump Pumps

No matter how long you have lived in your building—several months or several years—your basement is never really free from the possibility of water coming in from a heavy rain or melting snow. Not only is it upsetting to see all of your prized possessions floating in the basement, but in a situation such as this, your neighbors probably are in the same predicament, making it virtually impossible to get hold of a sump pump. If you are lucky, you will find one in a hardware store that is selling for about three times its usual cost. Having said all of this, the point is that you can never go wrong in investing in a standard floor-mounted pump that operates on a regular electric outlet and only needs a garden hose to discharge surface waters from your basement. Figure 7-32 shows this type of pump that usually costs approximately $50. As part of your building maintenance defense plan then, possessing such a pump makes great sense.

If your property is already equipped with a sump pump system, be sure to check it out on a regular basis. Every 3 months lift the float on the pump to run it. Make sure that there is water in the sump hole so you won't burn out the bearings in the pump. Before testing, make sure that there is no debris in the bottom of the sump hole; if there is, scoop it out before turning on the pump.

Sump pumps or floor-mounted pumps should be connected to a three-slot grounded outlet and never to an ungrounded outlet. If yours is plugged into some makeshift arrangement of wiring or into an older two-slot ungrounded outlet, be sure to call in an electrician to have a proper outlet installed. Improper outlets, electrical appliances such as pumps, and water all add up to serious shock potential.

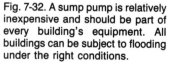
Fig. 7-32. A sump pump is relatively inexpensive and should be part of every building's equipment. All buildings can be subject to flooding under the right conditions.

If your pump makes an awful lot of noise when it is running, the bearings are probably worn out. Although they can be replaced, it is wiser and less expensive to just replace the pump. And while you are in a buying mood, it would be a good idea to buy a second pump to use as a backup pump just in case you need it in a crisis. This is particularly true if you have a history of wet basement problems.

Review the sump pump maintenance tips to recap what you should know about sump pumps and their maintenance. Remember, even if you have been fortunate in having a dry basement, there might come a day when you will need to know about wet basements—don't let it be at the last moment.

- Even with dry basements, have a pump available for emergencies.
- Test your sump pump every 3 months.
- Be sure that you have water in the sump hole when testing a sump pump.
- Periodically clean out any built-up debris in the sump hole.
- Make sure that your pump is connected to a grounded outlet.
- Loud noises may mean that the bearings in your pump are gone and a replacement pump is needed.
- Consider buying a backup pump just in case your main pump fails during a major flood.

Pipes and Drains

During your frequent basement forays, always be on the lookout for signs of deterioration in your waterlines, in waste lines, and in the drains and traps. Check for corrosion, deterioration, and signs of leaks. Gently tap on the cast-iron drains and waste lines and listen for sounds that might indicate weak and thin areas. Monitor spots in weak areas that you suspect might develop leaks. You might want to have your plumber look to confirm or disperse your suspicions. Always try to have repairs and replacements made before a broken waterline or a damaged drain floods your basement.

Shutoffs

Once a year test all the shutoff valves in your building by turning them off and then back on again. It is particularly important that you test the main shutoff (Fig. 7-33) to see if it can shut off all the water coming into your building. Test it by first shutting it off and then opening up a faucet in the building. If water continues to flow from the open faucet, then the main shutoff is defective and needs either repairs or replacement. Sometimes a simple solution is to have a plumber install a new secondary shutoff valve on the main line.

Fig. 7-33. As part of your maintenance program, annually check the operation of the main water shutoff valve in the building. Turn the handle all the way off and then run a faucet to see if the main shutoff is doing its job. Be sure to turn the handle all the way back on when you are done. Note the new electrical jumper grounding connection, which is becoming mandatory in most communities.

If you find that some of the shutoffs under fixtures are frozen (cannot be closed), don't try to force them as you very well might break the pipe or the valve. Instead, apply several drops of Marvel mystery oil (which can be purchased in any supply store) on the stem of the frozen shutoff. Do this several times until you are able to close the shutoff. If this fails to work, you might want to have the valve replaced.

Winterizing the System

If you close down your building for the winter, be sure to follow these suggestions for winterizing your system. These tips should prevent any freeze-ups while you are gone.

- Turn off all house water at the street shutoff valve.
- Open and drain all house faucets and outdoor spigots.
- Shut off the fuel to the hot-water tank, connect a garden hose to the drain, and flush the tank dry.
- Flush toilets and sponge out remaining water from the tank.
- Winterize all traps with antifreeze.
- Turn off fuel to the heating systems and drain the boiler.
- Drain radiators.
- Don't forget to drain appliances as well.

Frozen Pipes

Preventing your pipes from freezing up in the winter should be on your maintenance schedule if you live in an area where the temperature drops below freezing. Before old man winter comes calling, make a list of all the things to do. The following tips should help you get started:

- Insulate pipes and drains that are apt to freeze.
- Use electric heating cables on vulnerable pipes.
- Keep an open door between heated and unheated areas that have water pipes.
- If no insulation is available to you, try using several layers of newspaper wrapped loosely around pipes and tied with a string.
- On very cold nights, leave faucets dripping; this will prevent frozen pipes.
- Use 100-watt bulbs or heat lamps to warm walls with buried pipes.
- Repair broken windows near pipes and drains.
- Caulk and seal open joints that could allow in drafts.
- Install freezeproof outdoor faucets; be sure to turn the water off to outdoor faucets.

SUMMARY

Maintaining your mechanical systems (heating, cooling, electrical, and plumbing) requires continuous monitoring, checking, inspecting, and following preventive measures to keep your systems "healthy" and in good working order. As you continue to read and develop more understanding of what maintenance is all about, you will see that it will save you money in the long run. Good maintenance prevents failures and breakdowns that cost you money for repairs and replacements. Every so often read this chapter to recall what you should be doing to prolong the life of your mechanical systems.

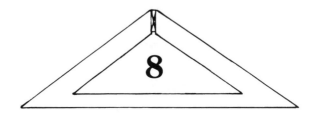

Hazards

In ordinary times, the headline, "Chlorodane Dangerous to Your Health," would probably receive some but not much attention and would soon be forgotten. These are not ordinary times, though; in fact, they are rather exasperating because more and more studies show that products once thought to be safe are now known to be hazardous to our health. Although it is not the intention of this chapter to scare you, it is essential that property owners are informed about such dangerous materials. Furthermore, they must also be made aware of other dangers that can be present in their homes because of faulty wiring, lead paint, or safety hazards, to name just a few. This chapter discusses all of these and other perils and provides timely advice, tips, and reference sources to help you identify, minimize, and—if possible—eliminate these hazards.

RADON

Radon is a radioactive gas (a by-product of uranium decay) that is released into the ground and that can seep into buildings from the foundations or basement floors. This unseen gas is particularly dangerous in enclosed structures, such as your building, and particularly so if the building is a well-insulated, tight structure. The Federal Protection Agency estimates that up to 20,000 people in the United States die each year from lung cancer caused by radon. A rather scary statistic!

Radon gas has been detected in dangerously high levels all over the country. Just because your neighbor's building tested out free of it does not mean that it isn't present in yours. Test results have varied greatly over distances as insignificant as 10 yards. There are estimates that approximately 12 percent of all American homes have higher than average levels of radon.

Because the gas is colorless and odorless, the only way you can tell the level in your building is by having a test done. To make sure, though, that this procedure is carried out correctly, contact only those companies that are recommended by the Environmental Protection Agency. You can call 1-800-334-8571 for a list of names, addresses, and telephone numbers of reputable firms that have passed the EPA's vigorous scrutiny. For additional information about radon and what to do about it, contact your local regional office of the EPA. Ask for the booklet entitled: *Radon Reduction Methods: A Homeowner's Guide.*

LEAD

Many older buildings throughout the country still have lead paint on walls and ceilings as well as some lead water pipes in the plumbing system. Occupants of such buildings might contract lead poisoning by drinking water that is contaminated with lead minerals from lead pipes or even from copper pipes with lead soldered joints. Inhaling lead paint dust while sanding wood that was painted with lead paint is another way of being poisoned. Paint chips ingested by children is probably one of the major childhood problems with long-lasting effects on the affected child. While a high concentration of lead in the body is quite toxic, lesser amounts can still cause severe learning disabilities in children, cause blood pressure and neurological ailments in adults, and might also cause complications during pregnancy.

Peeling paint in buildings built before 1978 (Fig. 8-1) causes many children to suffer lead paint poisoning. If you have children and your ceilings look like the one depicted

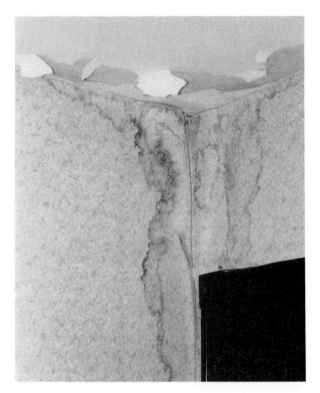

Fig. 8-1. Peeling paint in any building, but particularly older buildings, could signal lead paint hazard. Be sure to correct the source of the problem and then get rid of the peeling paint.

Fig. 8-2. Heat guns can help in removing lead paint in a relatively safe way. (Be sure to provide ample ventilation.)

in the photograph, scrape, prime, and paint the ceiling. Trying to delead an entire building or major portions of a structure can be very hazardous to you and the inhabitants. If you need to delead a portion of a room, try using a heat gun, as shown in Fig. 8-2, rather than sanding down the wood. The lead dust generated by the sanding can cause serious harm to anyone near it. On the other hand, using a heat gun according to the manufacturer's directions should cause little concern. Be sure to work in a well-ventilated work area.

There is a simple test you can do to determine if the paint in your building contains lead. Take sample chips from various areas and apply a drop of sodium sulfide solution to each sample chip. If the paint chip turns a dark gray or black, the chip contains lead paint. A local pharmacist, by the way, can supply you with the solution that you need. One word of caution: be careful in interpreting the results; dark-colored chips will not change color enough for you to tell if it is lead.

Old lead water pipes and/or lead solder on copper pipe joints poison the water in over 40 million American homes, according to the Environmental Protection Agency. If your building has lead pipes or joints, or if you suspect for any reason at all that you or your family members suffer because of your water, there are some steps that you can take to protect yourself.

1. Let the water run from the tap at least three minutes before you drink it or use it for cooking.

2. Flush the water out of your faucet if it has not been used for several hours.

3. Check your Yellow Pages for names of laboratories that test the quality of water. For a small fee technicians will analyze your water and tell you its lead content or anything else that is in it that should not be there.

4. If the lead content of your water is high, have your blood and that of your family tested to be on the safe side.

5. Next to removing the pipes and replacing them, the best thing you can do is buy and use bottled water.

For a copy of the booklet entitled: *Lead in Drinking Water: Should You Be Concerned?* write to EPA, Public Information (PM-211B), Garage Level, Southeast Mall, 401 M Street SW, Washington, DC 20460.

INSULATION MATERIALS

Insulation materials, although serving a useful function in a building, can actually cause more harm than good. Some materials are now known to be cancer-causing substances, while others have been known to cause fires in buildings. Let's look at some of the more common types and see what can be done about any problems that might exist in your building.

Asbestos

Insulation materials such as asbestos were used extensively for heating systems throughout the country from the 1920s to the early 1970s. Today, old boilers and their pipes still have this old dangerous covering in many basements. Asbestos is fine when it is left alone in its original condition; however, once deterioration sets in, fibers can easily float in the air and are very likely to irreparably harm anyone breathing them in.

How would you know that you have asbestos in your building? The photograph in Fig. 8-3 shows an example of asbestos insulation on a pipe, and Fig. 8-4 shows an old boiler covered with asbestos insulation. Sometimes this insulation has been removed from heating systems, and it is difficult to tell if it was ever there. You can see the residual fibers on a pipe that has had this material removed in Fig. 8-5. If you should have any concerns about asbestos contamination, consult an industrial hygienist who can test the air to determine the amount, if any, of free-floating fibers. Names of qualified industrial hygienists can be found in the local Yellow Pages. For information on asbestos content in consumer products, call the Asbestos Hotline at (800) 638-8333. If you can't find a testing laboratory near you, call (800) 424-8571 for assistance.

Make sure that you don't make the mistake of trying to remove the asbestos yourself. If you disturb these fibers, you could contaminate yourself and your entire building. Only a licensed asbestos abatement contractor should touch this stuff. It is really worth the

Fig. 8-3. Asbestos insulation on a heat pipe. Note the open end joints that could be a potential hazard. Open areas should be carefully wrapped and sealed.

extra cost to have a professional do the job. If you can't find a licensed contractor to do the work, here is a toll free number you can call for assistance: 1-800-424-9065.

Urea Formaldehyde

Urea formaldehyde is an insulating material that was used quite extensively during the late 1970s and the early 1980s. Prior to 1982, this foam insulation was pumped into buildings without much concern and regard for the inhabitants. Since that time, the Consumer Protection Safety Commission has found that improperly installed formaldehyde has serious side effects on the inhabitants of insulated buildings. Because so many people became ill from the gas that was emitted from the insulation, it was banned from use in this country.

In addition to the foam itself, other products in your building could also give off formaldehyde gases; for example, particleboard, plywood, and carpeting fall in this category. If any of these materials were used in your building, refer to Table 8-1 to see what you can do about alleviating the problem. Additional information can be obtained from the Consumer Federation of America, 1314 14th St., NW, Washington, DC 20005. They have a very useful publication entitled: *Formaldehyde: Everything You Wanted To Know But Were Afraid to Ask.*

Fig. 8-4. Asbestos on an old coal-converted boiler. The boiler is on its last legs, and when replacement time comes, special safety precautions must be taken to protect the home from free-floating asbestos fibers.

Fiberglass Insulation

Is anything safe? Recent studies have shown that even common fiberglass insulation, with which most new homes and many older ones are insulated, is a possible cause of lung cancer. Again, as is true with so many other synthetic materials, there is a link between lung cancer and exposure to relatively low levels of airborne fibers. Ironically, fiberglass was touted as the safe replacement for other hazardous insulation substances such as asbestos and formaldehyde.

Not all is lost, however. There are some steps that you can take to minimize and prevent the inhalation of such dangerous fibers. If you have business to do in an attic that is insulated with fiberglass, for example, always wear a face mask (Fig. 8-6). If you have insulation that is not covered and sealed, such as in a basement, consider applying Sheetrock or a plastic covering to eliminate the chance of floating fibers. And always be sure to read the directions for installation given by the manufacturer.

Fig. 8-5. Heat pipes with their asbestos insulation removed. Note the residual fibers still clinging to the pipes and joints. A basement that has had asbestos removed should have an industrial hygienist test the air for active fibers.

Table 8-1. Sources of Formaldehyde Fumes.

Source	What to Do
Urea formaldehyde foam insulation	Remove insulation—very costly. Provide ample ventilation in building. Caulk and seal all cracks and joints in walls and where walls meet floors. Add an air-to-air heat exchanger.
Particleboard	Apply a particleboard sealer. Paint or varnish exposed areas. Install vinyl wallpaper and floor covers. Provide ample ventilation in building.
Plywood	Apply sealants, paints, and varnish. Ventilate building.
Carpeting	Shampoo carpets to dilute formaldehyde. Ventilate and use a dehumidifier.

Fig. 8-6. Be sure to wear a mask when working with fiberglass insulation.

Insulation and Fire Safety

Although the recorded number of fires related to insulation materials is small compared to the millions of homes that are insulated each year, you most likely don't want to be included in such a statistic. Let's look at the causes of fires that started in insulation materials.

Some insulation materials used in the past, such as cellulose, were not treated chemically with flame retardants. Many a house fire started in such untreated combustible material. If your building was insulated before 1978 with cellulose, there might be a chance that it was not made fireproof. To test it, here is what to do: Take a handful of cellulose insulation and drop it in a sink basin. Then put a match to it to see if it ignites. If it does, you might want to call a contractor to have it double-checked and probably removed. Additional advice can be obtained from the National Association of Home Insulation Contractors (NAHIC). Either write to them at 4750 Wisconsin Ave., Washington, DC 20016, or call (202) 363-8100.

Apart from the possibility of the insulation itself catching fire, the following are the most common causes of insulation-related fires:

1. Insulation was placed over heat-producing devices. The heat gets trapped and builds up to the ignition point of the materials covered by the insulation. It is therefore very important that you never cover heat-producing devices with insulation. The following are some examples:

- Recessed light fixtures
- Chimney flue pipes

190

- Chimneys
- Exhaust motors and fans
- Heat pipes or heat ducts
- Exposed electrical wiring
- Open junction boxes
- Doorbell transformers

2. Insulation materials were installed incorrectly. Some insulation materials have backings that are highly flammable, such as batts with paper facing. These should not be installed upside down where exposure to an accidental spark or a carelessly discarded cigarette could ignite it.

3. Heat-producing gadgets used in insulated areas can cause fires. Always make sure that you and others working in or around insulated areas don't get careless.

- Keep cigarettes and other smoking materials away from insulation.
- Keep droplights away from flammable materials.
- Don't use torches around insulation.
- Don't use matches in lieu of a flashlight.

4. Carelessness on the part of tenants and property owners can cause fires. Here are three of the most common careless acts that can cause insulation fires:

- Using light bulbs that exceed the manufacturer's rating for a particular light fixture
- Stuffing insulation into recessed light fixtures to block off air drafts
- Overloading or overfusing the building's electrical circuits, which might cause wiring to overheat. Adjacent insulation will then focus this heat to nearby flammable materials

5. So-called "Acts of God," such as lightning and fires that spread from one building to another, are things that are out of your control. Concentrate on the first four causes to help minimize your chance of a fire and leave the last to luck.

Probably the single most common cause of insulation-related fires, according to the National Association of Home Insulation Contractors, is the overheating of recessed light fixtures. The best way to avoid such a calamity in your building is by following the directions given in Fig. 8-7. Keep all insulation a minimum of 3 inches away from recessed light fixtures and never place insulation or other materials directly over the fixtures. If you have lights that look like what you see in Fig. 8-8, then you have some recessed lights. If that is the case, be sure to check the areas above them to make sure that there is no chance of a fire. Remember that insulation is supposed to keep heat where it belongs and not to heat up flammable materials.

FIRE HAZARDS

Is it possible that your building is a fire trap? If you are not sure of the answer, just read on. Every year, 300,000 homes are destroyed by fires, and as many as 4,000

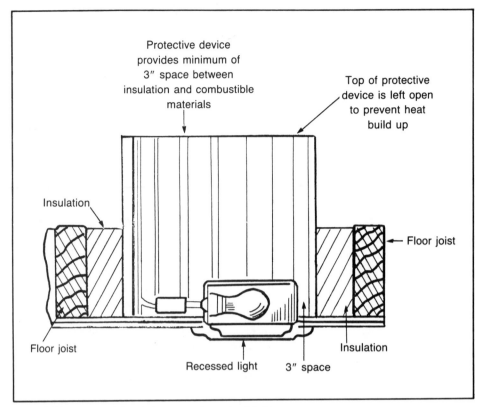

Protective device provides minimum of 3″ space between insulation and combustible materials

Top of protective device is left open to prevent heat build up

Insulation

Floor joist

Floor joist

Insulation

Recessed light

3″ space

Fig. 8-7. How to install insulation around recessed lights.

Fig. 8-8. Remember not to cover recessed lights with insulation.

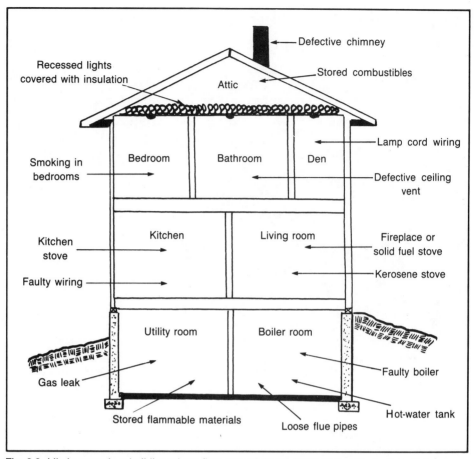

Fig. 8-9. Likely areas in a building where fires can originate. Make sure that you keep these areas clean and free of any fire-producing materials.

people are killed because of faulty heating systems, defective wiring, malfunctioning chimneys and fireplaces, and careless storage of flammable liquids and materials.

Because you don't want to add to this number, make sure that you take an inspection tour just for the purpose of finding out if you have any fire hazards in your building. If you detect the slightest potential, correct it immediately. To say, "I am going to do it tomorrow," is just not good enough! And always be prepared to cope in the event of a fire—you should have a working escape plan in effect. Figure 8-9 illustrates the areas in a building in which fire is most likely to start. The National Safety Council has a number of useful publications on fire prevention that might be of some help to you. Call (312) 527-4800 for information.

Heating Systems

A heating system that is not properly serviced, cleaned, and inspected every year could easily cause a fire. Give your heating system top billing in your maintenance program. Check flue pipes for weak spots and make sure that it is properly supported every few

Fig. 8-10. Heat-producing appliances, such as hot-water tanks and heating plants, need to have flue pipes connected to and sealed firmly into chimneys. Open areas, as seen in this photograph, can prove to be a disaster.

feet. In Fig 8-10 you see a flue pipe just waiting for a fire to start. Make sure you make the necessary repairs to anything that looks like that. Also, be sure to move all combustibles or flammable materials away from your heating system, particularly flues and burners. Make it a habit to keep all combustibles (wood, newspapers, trash, paint, rags, etc.) away from the boiler room.

Annual tune-ups should be part of a regular preventive maintenance program that your fuel company should provide. And pay particular attention to solid fuel stove maintenance and cleaning. Excess creosote, as seen in Fig 8-11, can be the catalyst that burns down your building. As with some maintenance chores, it is always wise to have a professional clean and tuneup your heating system. For the phone number of a professional chimney sweep in your area call (202) 857-1181 or write to Wood Heating Education and Research Foundation (WHERF), 1101 Connecticut Avenue, N.W., Suite 700, Washington, DC 20036.

Chimneys

Faulty chimneys can cause a major fire that could destroy your entire building. With this in mind, be sure to go over safety check points very carefully and do it every year. Check the height of the chimney on the outside. On a pitched roof it should be 3 feet above the roofline and at least 2 feet higher than any combustible materials within 10 feet. See the 3-2-10 rule as illustrated in Fig. 8-12. If your chimney does not measure

Fig. 8-11. Keep chimney cleanouts and flue liners free of combustible creosote buildups.

up to these standards, ask a local mason to make the necessary corrections. While the mason is at it, have him also check and make necessary repairs to the cap, brick joints, and flashing of the chimney.

It is important not only to check the outside of your chimney but to inspect the attic areas as well. Repair loose, cracked, and deteriorated brick, stucco, and mortar immediately. Figure 8-13 shows a chimney in an attic that is begging for a fire.

Seal unused flue openings (Fig. 8-14) with solid masonry, not a metal cap as shown in this photograph. Caps rust out in time and will allow flue gases into the building, and in case of a fire in the chimney, these caps could blow out and allow the fire into the building.

Fireplaces

Common sense is all that you need in working with fireplaces. Always use a spark screen or a heat-resisting glass door (Fig. 8-15) to avoid sparks igniting your carpeting. Never have combustibles near the hearth, particularly if you have carpets. And never leave a fire unattended—unless you like living dangerously. It doesn't take long for flames to start licking their hot tongues up the walls of a room. Before you retire for the evening, be sure that the fire is out and close the glass doors on your enclosure, if you have one. At least once a year, check the condition of the hearth, the fire walls, the damper, and all visible areas of the fireplace. Always follow directions when you are using artificial logs because they burn more intensely in comparison to wood logs. Never use more than one artificial log at a time. Be sure to have your chimney flues inspected and cleaned on an annual basis.

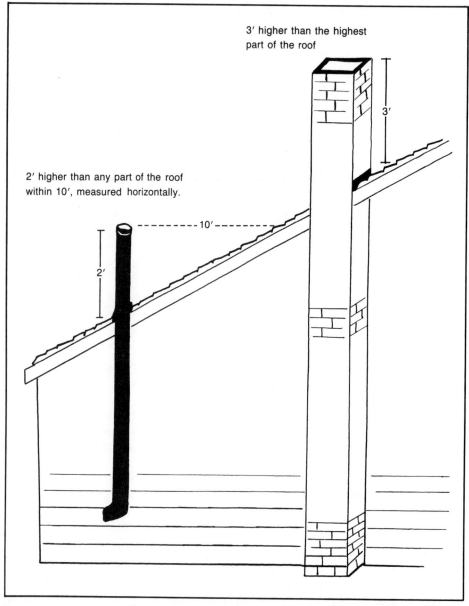

3' higher than the highest part of the roof

3'

2' higher than any part of the roof within 10', measured horizontally.

10'

2'

Fig. 8-12. The 3-2-10 foot rule for chimneys. Check your chimney to make sure that it complies with this rule.

Space Heaters

If you have to use a space heater, be sure that it is one that bears the Underwriters Laboratory (UL) tag. In addition, gas space heaters should also have an approval by the American Gas Association (AGA) on the heater. If you are in the market for a space heater, be sure that the one you buy has these approval seals on them. It means that the heater has been tested and found to meet industry standards.

Fig. 8-13. Open holes in chimneys and deteriorated mortar joints and stucco finish is a potential problem that must be corrected. Practice your cement skills and seal up any such deficiencies in your own chimney.

Some states do not allow unvented space heaters, such as kerosene heaters. Be sure to check with your local building department if you plan on using a kerosene heater. A strong word of caution is given in regards to any unvented heater—*don't use them*!

As is true with any heating device, be sure to maintain safe clearances from combustibles and only use a space heater in accordance with the manufacturer's recommendations. Be sure to keep children away from them at all times. Never substitute a different fuel from the one recommended. Never fill a hot kerosene space heater with fuel, unless you enjoy fireworks. Remember that gas and oil heaters need to be vented to a chimney to vent toxic combustion fumes to the outdoors.

Make sure that electric space heaters are connected to outlets that can carry the electrical demand of the heater. If in doubt, check the label on the heater or read the manufacturer's directions. Never use makeshift wiring, such as lamp cord attachments, to provide power for electric space heaters.

Electrical Appliances

Electricity used improperly can be a major contributor to fire hazards in a building. Careless use of appliances, damaged electric cords, overfused circuits, and old and inadequate wiring account for most of the building fires in this country.

Fig. 8-14. Metal caps are not considered safe chimney hole seals. Use brick and mortar to cement up and seal any unused opening in a chimney.

Remember that your dryer, your stove, your air conditioner, and all other such major appliances need separate circuits and that each one requires an adequately sized fuse or circuit breaker. Table 8-2 lists some of the more common appliances with the correct fuse or circuit sizes.

Never double up a toaster, a radio, and an iron, for example, on one circuit (Fig. 8-16). This foolishness is bound to result in some serious problems. If you don't have enough outlets, have an electrician add more circuits and outlets, particularly in heavy-use areas like the kitchen.

Don't block air circulation around TV sets, stereos, computers, and refrigerator compressors. By putting such appliances in enclosed areas or by placing obstructions to air near them, you provide a climate for a fire.

Never walk away from heat-producing appliances such as a hot iron. When you are through with an iron or a toaster, be sure to unplug them to avoid any accidents. Never put a hot iron away until it is cold.

By maintaining and keeping appliances clean and in good working order, you run less risk of an accident. Repair or replace damaged or frayed electric cords. Never

Fig. 8-15. Always have some form of spark screen in front of a working fireplace. It doesn't take much to ignite a wood floor or a rug.

Table 8-2. Appliance Circuits.

Appliance	Fuse or Circuit Breaker Size
Electric stove	50 amp
Water heater	30 amp
Dryer	30 amp
Disposal	20 amp
Dishwasher	20 amp
Washing machine	20 amp
Freezer	20 amp
Refrigerator	15 amp
Microwave oven	15 amp

It is customary to provide a separate circuit for each of the listed appliances.

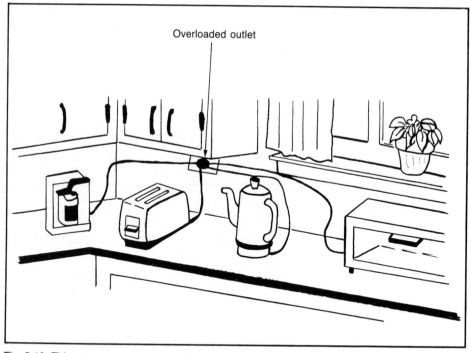

Fig. 8-16. This octopus arrangement will eventually lead to a fire. Use common sense and have an electrician add as many new outlets as you need to make your life easier and safer.

overload electrical cords. Never run cords under rugs, behind radiators, along sharp metal objects, and *never* up through a heat duct (as seen in the photograph of a heat register in Fig. 8-17). If you find any such defects in your building, make immediate corrections.

Electrical Wiring

If you are busy replacing fuses all the time and resetting circuit breakers, you should realize that your electrical system is trying to tell you something. You probably need more circuits and perhaps an updated electric service. Have an electrician evaluate your present system to see what can be done about alleviating the problem.

If you find yourself replacing fuses, be sure to always use the same size fuse that you removed. Never replace a blown fuse with one of a higher rating; for example, never substitute a 30-amp fuse for a blown 15- or 20-amp fuse. Above all, *never* substitute a copper penny for a fuse. That one penny could cost you your building. Overheated wires will not be able to break the circuit, and the results could be a major fire.

If you suspect that the wiring in your building no longer meets your needs, have your electrician check out the system as soon as possible. Some tell-tale signs of an inadequate system are:

- Constant blowing fuses or tripping circuit breakers
- Appliances that heat up slowly
- Dimming lights
- Shrinking TV pictures

Fig. 8-17. Don't take the easy way out. Only a fool would run electrical wiring up through a heat duct. Not only will the heat affect the wiring insulation, but there is always the chance of the metal cutting through the live wire. Always have a licensed electrician make major repairs and renovations.

- Warm outlet covers
- Mysterious sounds and strange odors coming from outlets and switches

Good Housekeeping

Good housekeeping habits—indoors and outdoors—not only give you a sense that everything is under control, but provide a certain degree of safety around the building. Always store flammable substances and materials such as paints, lubricants, fuel, and cleaning materials in tightly sealed metal cans, preferably in a shed away from your building. Never store gasoline or gas-fueled appliances or tools in your building, such as a lawnmower. Also get rid of trash and debris before it builds up and gets out of hand. Pitch oily rags right after you are done with them. Clean up sawdust and wood scraps whether you are working indoors or outdoors. Picking up gives your building a tidy look as well as making it safer.

When it is time for a family cookout, keep charcoal fires away from the building. Never use kerosene or lighter fluid to start up a fire because you might just barbecue your home instead of the dinner. Buy yourself a few good quality fire extinguishers and store them in key locations, again both indoors and outdoors. Know how to handle them properly in case you need to use them.

Flammable Liquids

Gasoline, kerosene, acetone, and paint thinners are some of the liquids that give off vapors that could easily ignite, or worse, explode—so treat them with the utmost respect. In Fig. 8-18 you see a garage where the owner is obviously oblivious to the danger. Limit your supply and keep them on hand only if absolutely necessary. Store all these items in airtight cans if you can and in cool, well-ventilated areas. It is always helpful if things are properly labeled, and these liquids are no exception. It saves you opening and closing lids unnecessarily. Be sure to always wipe up spills immediately and pitch those rags when done. *Never* smoke around flammable liquids as the slightest spark could cause a flash fire or explosion.

Fig. 8-18. Never store flammable liquids in a garage. Follow the directions given in the text for safe storage.

Smoking and Matches

Needless risks are taken daily by millions of Americans—not only to their lungs but also to their homes—by the habit of smoking. If you are a smoker, why not take this time to do both your lungs and your building a favor and quit. At least follow these guidelines to avoid the slightest chance of starting a fire.

- Use ashtrays that are noncombustible and are large and deep.
- Before you dispose of cigarette butts, douse them with water first and then pitch them into your wastebasket.
- Don't be careless and drop ashes on rugs or upholstery where they could smolder without anyone noticing it.
- Always close matchbook covers after using them.
- Store matches out of reach of children.
- Be careful how you discard matches; make sure that they are out.
- Never use matches in place of a flashlight.
- Teach children a ''healthy'' respect for matches.
- Don't leave a lighter where a child could get it.
- Never rest a lit cigarette on flammable materials.

Develop a Safety Escape Plan

Even though you have tried to provide the best inspection and maintenance for your building, you might still have some fire hazards. Planning ahead will give you that needed edge if the worst scenario ever comes about. Your building or home should have a safe escape route that all the inhabitants know. You should also have a fire drill periodically to review your escape plan. Bedrooms on upper levels should have a rope or metal escape ladder handy that can be used if people are trapped. Windows in all bedrooms should be easy to open for emergency escapes. Smoke detectors, both the battery type and those that are tied into your electrical system, should be in place and functional. Your local fire department will be more than happy to show you where to install them. Usually one detector is placed on each level of the building and outside of bedrooms as well. Check your smoke detectors as often as recommended by the manufacturer. Never stay in a burning building to salvage some material goods—the goods can be replaced, but you cannot.

CARBON MONOXIDE

Whether you heat your building with gas, oil, coal, or wood, your heating system needs fresh air to work properly. If your boiler or furnace does not get all of the oxygen that it requires, it could start producing carbon monoxide—a colorless, tasteless, and odorless gas that is very toxic. Exposure to it causes headaches, dizziness, weakness, nausea, or loss of muscle control. Prolonged exposure to high concentrations can lead to unconciousness, brain damage, or even death. Because you want to avoid any of the noted conditions, learn how it occurs and what you can do about it.

If bird's nests, leaves, or tree branches block your chimney, or if considerable amounts of soot have built up to block your chimney, you are running a high risk of generating carbon monoxide in your building. Figure 8-19 illustrates several causes for

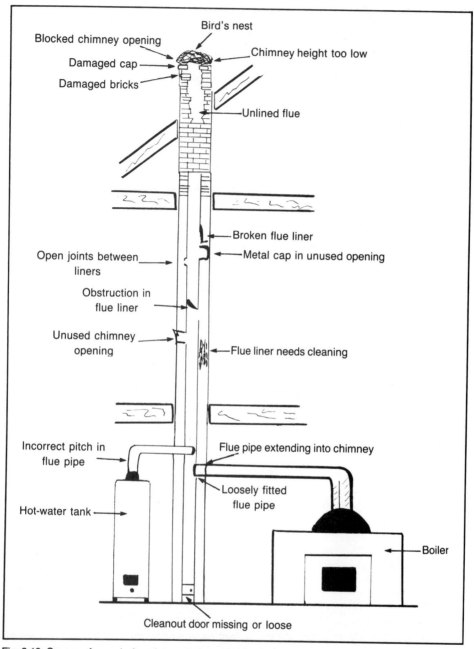

Fig. 8-19. Causes of poor draft performance in a chimney. Have a chimney sweep inspect your chimney to make sure none of these defects presently exist.

a blocked chimney. If you suspect carbon monoxide in your building, call in a chimney sweep to check your chimney. If there is nothing wrong with the chimney, have your heating system tuned up. For immediate first aid for carbon monoxide poisoning, provide lots of fresh air and get medical attention as soon as possible.

TREATED WOOD

Although pressure-treated wood is great to use for outdoor projects such as decks and porches,' it does still pose a hazard to you. Pressure-treated wood contains inorganic arsenic that helps to preserve it and keeps insects from eating it. This same poison that gives it many years of useful life could shorten your life, so pay particular attention to the following guidelines set down by the Environmental Protection Agency.

Handling Tips:

- Dispose of treated wood by ordinary trash collection.
- Do not burn treated wood in fireplaces or wood stoves.
- Avoid inhaling the dust from sawing or sanding.
- Wear gloves when handling it and wash exposed areas of your skin before you eat, drink, or smoke.
- Launder all clothes that were exposed to the wood or to the sawdust separately from other household clothing.
- Don't make countertops, carving boards, or cabinets from treated wood.
- Avoid any contact of food with treated wood.
- Keep wood out of contact with drinking water.
- Don't build sandboxes or playpens with treated wood.
- See a doctor if you get a splinter from treated wood because it might poison the skin tissue.

HAZARDOUS HOUSEHOLD WASTES

You would be surprised at the amount of hazardous materials your neighbors are storing in their garages and basements. The shock only gets worse, however, when you realize that you are a collector of hazardous goods as well, perhaps without realizing how dangerous they really are.

Hazardous substances include corrosive, reactive, ignitable, and toxic materials such as oil-based paints, motor oil, pesticides, batteries, cleaning agents, and antifreeze products. Look around your building—chances are you store most or all of them. Now the question is what to do with these substances. Please don't pour them down drains, flush them down toilets, or dump them into the garbage and trash because you will only add to the contamination of our air and water.

Hazardous liquids that are poured down drains or toilets easily corrode the plumbing drains and, in time, might release fumes from plumbing traps that could filter back into the building. On a larger scale, these liquids might also cause damage to a septic or cesspool system or to a city water-treatment plant. Furthermore, hazardous liquids could leach into the ground water and pollute it.

Hazardous wastes put into the household trash and taken to landfill dump sites or incinerators will most likely end up contaminating the air and ground, polluting the ground water, and, in the long run, making its way back up the ecological chain to the polluters—humans.

There are many things that you as an individual can do to lessen the environmental impact on your property, your neighbor's property, and the environment in general. First, find out if your community or state has a hazardous waste collection site. If so, take your hazardous materials there or have a licensed transporter collect and dispose of them

**Table 8-3. Substitutes for
Commercial Household Products.**

Instead of Using	Try
Air freshener	Set vinegar out in an open dish. Place orange or lemon peelings on a warm stove top.
Drain cleaner	Pour boiling water down the drain. Use a plunger or plumber's snake to clear the line.
Furniture polish	1 teaspoon lemon oil in 1 pint mineral oil.
Insecticides	Soapy water on leaves. Natural organic sprays such as pyrethrum.
Moth balls	Put clothes in cedar closets or chests.
Oven cleaner	Salt, baking soda, and water mixed with elbow grease.
Roach spray	Boric acid (might be harmful to animals and children).
Silver cleaner	Soak silver in 1 quart warm water containing 1 teaspoon baking soda and 1 teaspoon salt, with a piece of aluminum foil.
Toilet bowl cleaner	½ cup of bleach.
Window cleaner	2 tablespoons vinegar in 1 quart water. Rub with newspaper or leather cloth.

for you. Secondly, you can substitute other substances for commercially available household products. They might not work as easily or as effectively, but they do help reduce the number of chemical products that you have in your building. Table 8-3 gives a list of alternative products that you might wish to try. Finally, follow the list of do's and don'ts for careful use of hazardous household products.

Dos and Don'ts:

• Properly label all hazardous materials and keep them separated from each other as well as other household goods.
• Always follow directions on the labels for usage.
• Avoid all contact with the skin and eyes.
• Keep children and pets away from these materials.
• Take your used motor oil to a local auto service station for recycling.
• Use alternative household products that are less hazardous.
• Only purchase small quantities of necessary products so that you will not have any leftovers.
• Rinse empty pesticide containers three times before you pitch them; dilute the water that was used before pouring it in your backyard.

- Don't mix hazardous materials.
- Don't flush wastes down drains or toilets.
- Do not bury hazardous wastes in your backyard or in the woods.
- Don't pour waste products down public sewer drains.
- Don't discard hazardous wastes in your trash or rubbish.
- Don't discharge hazardous wastes into waterways (lakes and rivers).
- Don't rely on chemicals to do the work that you should be doing, such as pulling weeds rather than spraying them with poison.
- *Use common sense.*

POISONPROOFING

To make your home safe from toxic substances, follow the helpful tips listed below. Table 8-4 gives some simple hints to do before medical help arrives. Buy yourself a good first aid book and read it.

- Use safety latches on cabinet doors, as shown in Fig. 8-20. These prevent small children from opening a Pandora's box.

Table 8-4. First Aid Chart.

Poisonous Substances	Treatment
Household cleaning, polishing, and painting agents	Give glass of water or milk. *Do not* induce vomiting.
Acids and alkali	Give glass of water or milk. *Do not* induce vomiting.
Medicines	Give glass of water. Induce vomiting.
Gasoline and petroleum products	Give glass of water or milk. *Do not* induce vomiting.
Food poisoning	Give glass of water. Induce vomiting.
Insect and rodent poisons	Give glass of water. Induce vomiting.
Poisons in the eye	Flood injured eye with water.
Poison on the skin	Remove clothing and flood the involved body parts with water.
Poisons inhaled	Move person to ventilated area. Resuscitate (mouth to mouth) if person has stopped breathing.

Fig. 8-20. Safety latches on cabinet doors prevent children from getting at dangerous substances.

- Do not rely on childproof medicine containers to ensure safety.
- Keep all medicines and prescriptions out of children's reach.
- Don't refer to medicine as candy; it will make the medicine more tempting to a child.
- Remember, anything you use to clean, polish, open drains, or kill pests with can be poisonous. Keep such products out of children's reach.
- Don't transfer one product to the container of another one.
- Keep *syrup of ipecac* on hand at all times (you can buy it at your local pharmacy). Use it to induce vomiting.
- If in doubt about an emergency, call the Poison Center Emergency Hotline (301) 722-6677.
- Don't induce vomiting when corrosives or petroleum products have been swallowed. It will cause more damage (see Table 8-4).
- Always maintain respiration and circulation and administer first aid treatment while waiting for medical help.
- Remember, quick action is extremely important in cases of accidental poisoning.
- Learn first aid and become familiar with first aid procedures.

BURGLARPROOFING

To prevent your home, apartment, or business from becoming an easy target for thieves or another crime statistic, it is always wise to do as much as possible to avert an unwanted entry. To help you make your property safe, the Insurance Information

208

Institute has published a brochure entitled, *Home Security Basics*, which provides a host of suggestions. For a free copy, call the toll-free consumer hotline 1-800-221-4954, or if you prefer, write to the Institute at 27 School Street, Suite 305, Boston, MA 02108.

The following burglarproofing guidelines should help you for now:

- Ask a neighbor to keep an eye on your home while you are at work.
- Have an electrician install outdoor lighting. (For information about outdoor lighting, write to: GE Inquiry Bureau, Nela Park, Cleveland, OH 44122; ask for *The Light Book*.)
- Leave window shades up and keep a radio playing to give the impression of someone being home.
- Equip all entrance doors with dead-bolt or double-cylinder locks.
- A slide bolt with a key will help secure a sliding glass door.
- Don't leave extra keys around the mailbox, doormat, or under flowerpots.
- Change locks if keys are lost or stolen.
- Never leave your house keys with a parking lot attendant.
- If your property is in a high-crime area or if you have a lot of valuables in your home, consider installing a burglar alarm.
- Have safety locks (Fig. 8-21) installed on windows.

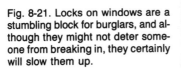

Fig. 8-21. Locks on windows are a stumbling block for burglars, and although they might not deter someone from breaking in, they certainly will slow them up.

- Make sure that valuables, such as paintings and antiques, are not visible from the outside.
- Use simple delaying tactics such as padlocks, door and window locks, grates, bars, and bolts to discourage intruders.
- Never go out and leave windows or doors unlocked.
- If you see signs of a break-in at your property, don't go in. Call the police from a neighbor's home.
- Before you go away on vacation or leave your home or business for any length of time, be sure to notify the postal authorities to put a hold on your mail delivery—an overflowing mailbox is an open invitation to a burglar.

INDOOR POLLUTION

Once you have made your home burglarproof and safe from poisons and hazards, you should feel pretty well protected. And if your home is nice and tight—in other words is well caulked and weatherstripped—you should also feel safe from the elements with a relatively reasonable satisfaction. But you can't! There still is danger lurking in your building. It is silent, virtually odorless, colorless, and you would not be able to detect it unless you happen to have specific scientific equipment, which is unlikely. The illustration in Fig. 8-22 shows how indoor pollution can harm you and your family members. Exposure to indoor pollutants over a period of time is often the cause of chronic headaches, impaired lung functioning, cancer, and even death by asphyxiation.

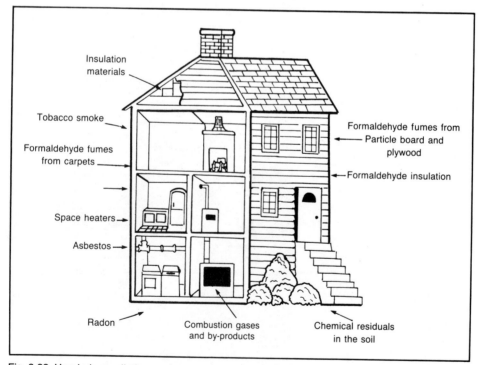

Fig. 8-22. How indoor pollution can harm you and your family members. Take all the necessary steps to avoid indoor pollution.

The major culprits of indoor pollution can be categorized as follows:

1. *Combustion by-products and gases*—include carbon monoxide, carbon dioxide, and other by-products that are released by poorly vented furnaces and space heaters as well as the pervasive indoor pollutant of cigarette smoke.

2. *Biological substances*—include bacteria that breed in cooling systems, fungi growth on damp air-conditioner filters, and mites that grow on dust particles. In addition, allergy-producing substances such as pollen can be brought in from the outside by ventilation systems.

3. *Chemical residues*—pesticides that are sprayed on walls and floors might give off small amounts of vapor for as long as months and sometimes even years. Formaldehyde (as already discussed) that is used in insulation and carpeting as well as to make plywood and particleboard also gives off dangerous fumes and gases.

You can control indoor pollution in two ways: isolate or remove polluting agents or sources from your building and increase ventilation to provide circulating clean air. If you decide that your building is too polluted, here is a series of steps that you can follow:

Building structures:

- Seal all cracks in basement walls and floors with polymeric caulks, epoxy paints, and polyethylene or polyamide film vapor barriers to counteract the introduction of radon from the surrounding soils.
- Ventilate attics and crawl spaces.
- Have air-to-air heat exchangers installed to flush stale air out and to warm incoming fresh air to save on heating bills. Table 8-5 lists the names of several manufacturers who will be glad to send you their specifications.

Bathrooms:

- Ventilate moisture to the exterior via ceiling or wall-mounted fans.
- Keep toilets, tubs, and sinks free of mold.
- Dry walls and floors to discourage microbial growth on damp surfaces.
- Clean off mold and mildew with a strong solution of vinegar, hot water, and bleach.

Kitchens:

- Use a stove vent hood when you are cooking.
- Never use your cooking stoves to heat your home.
- Gas combination stove and heaters need to be vented to a chimney.

Cleaning products:

- Limit the use of aerosol spray cans.
- Use cleaning substances and techniques that don't pollute.

Heating systems:

- Have your heating system annually cleaned, inspected, and serviced.
- Make necessary repairs to chimneys and flues.
- Replace old, defective, and inefficient burners.

Table 8-5. Manufacturers of Air-to-Air Heat Exchangers.

Wood stoves:

- Buy a stove with a catalytic converter to reduce the emission of hydrocarbons.
- Only burn well-seasoned dry wood.
- Never burn wet or green lumber.
- Never burn pressure-treated wood.

Humidifiers:

- Clean with a strong solution of vinegar and hot water daily.
- Don't allow water to stand for long periods of time; this promotes fungi growth.
- Try to start up your humidifier with hot water because it contains fewer microorganisms.

212

Air conditioners:

- Clean coils and inspect for microbial growth.
- Have system inspected and serviced on annual basis.

Tobacco smoke:

- Limit or prohibit the smoking of tobacco products in your building.
- Restrict the smoking of tobacco products to well-ventilated rooms.

Space heaters:

- Use only UL-approved heaters.
- Vent space heaters to the outdoors.

House cleaning:

- Don't use aerosol spray cans for dusting, polishing, or spot-cleaning.
- Brush and groom pets outdoors.
- Use silicone-treated dustcloths to hold the dust.

ELECTRICAL SAFETY

You have already seen how fires can start because of faulty electrical wiring. Now let's see how electricity can literally shock you to death. Data collected by the U.S. Consumer Product Safety Commission show that over 600 persons each year lose their lives by electrocuting themselves in or around buildings. The information that follows should make you more watchful of electricity and should show you how to avoid a possible tragedy.

Most electrocutions in or around buildings happen because people either inadvertently contact power lines, use electrical appliances in or around water, or use electrical appliances that are installed incorrectly or were not properly repaired.

If you hit a power line with a metal ladder, an antenna, or a metal pole, for example, you would most likely suffer a tremendous jolt or be electrocuted. Follow these safety guidelines when you work near a power line:

- Use a wood ladder rather than a metal one.
- Never place your ladder where it could slide into a power line.
- When you carry or move a ladder, lower it sufficiently so you won't run into any power lines.
- When working on a ladder, always be aware of where you are in relation to power lines.
- Never do work standing on a ladder during heavy winds or rain.
- Install antennas at a safe distance from power lines.
- When you take down or put up an antenna, be sure to have someone help you.
- Never install or do work on an antenna during wet or windy weather conditions.

Besides staying away from power lines, always remember that the combination of water and electricity can be absolutely *deadly*. Never handle anything that has the slightest thing to do with electricity when you are standing in a puddle or even damp areas. With

Fig. 8-23. Consider adding ground fault outlets to all bathrooms, kitchen outlets near the sink, outdoor receptacles, garage areas, workshop areas, and swimming pool circuits.

water in the picture, an otherwise slight shock can become lethal. Many accidents have and will occur around the house while using hair dryers, power tools, radios, TV sets, sump pumps, and kitchen appliances simply because people are careless or do not read the directions.

Many accidental shocks could also have been averted and many lives saved if the electrical products had been plugged into a special safety device called a *ground fault circuit interrupter (GFCI)*. Figure 8-23 shows you what this outlet looks like. Today, the National Electric Code (NEC) calls for these outlets to be installed in kitchens, bathrooms, garages, outdoor outlets, and swimming pools because they can detect, within a fraction of a second, an electrical fault and shut off the supply of electricity before serious injury or death can occur. If your building does not have such lifesaving outlets, be sure to consult with an electrician about having them installed in appropriate locations. Meanwhile, follow these safety guidelines:

- Don't leave your hair dryer plugged in when you are not using it. If if falls into your sink or bathtub, it can still kill you, even though it's switched off.
- If you have children around the house, make it a rule that all electrical products are unplugged when not in use.
- Never dry your hair with a hair dryer while standing on a wet bathroom floor.
- Don't place radios or TV sets near bathtubs or sinks.
- Install ground fault circuit interrupters in all bathrooms, kitchens, garages, outdoor outlets, and pools.
- Repair or discard all appliances that cause even the slightest electrical shock.
- Never touch an electrical appliance and a ground at the same time. For example, a faucet is a ground. So, don't handle electrical appliances at the same time you have one of your hands on the faucet.

Repairing an electrical fixture or appliance or adding an electrical circuit yourself is, at best, a risky business. You really should leave such jobs for your licensed electrician. You might want to do some simple chores yourself, however; so with this in mind, be sure to follow these guidelines:

- Never leave an electrical appliance plugged in while you are repairing it.
- When you work on an outlet or a light fixture, be sure to pull the fuse or trip the circuit breaker for that particular circuit.
- Whenever you de-energize a circuit, make sure that no one energizes it while you are working on that line. It is a good idea to leave a note on the service panel explaining what you are doing.
- If your appliance or electrical product is old and has some mileage on it, consider just replacing it rather than trying to repair it.

TRIPPING/FALLING HAZARDS

Missing handrails, whether they are on the outside (Fig. 8-24) or on the inside (Fig. 8-25), pose a major potential hazard. Also, damaged railings (Fig. 8-26) or railings with missing spindles (Fig. 8-27) are almost as bad as having none. If your building is missing handrails or has damaged sections as illustrated, be sure to have the necessary installation or repairs made before someone is injured.

Additional tripping or falling hazards are present where you have uneven sets of steps or if the steps are too high. A normal range for risers can be in the 7-inch to 8-inch

Fig. 8-24. Any set of stairs outdoors with three or more steps (risers) should have at least one safety handrail.

Fig. 8-25. All indoor stairways should have at least one handrail.

range. The step in Fig. 8-28 is too high—over 12 inches. If you have a building that has uneven step risers or risers higher than normal, it would be twice as wise to make sure that these steps have handrails that someone could grab onto in case of an accident.

Sometimes people start projects and along the way somehow forget to complete them. As you can see in Fig. 8-29, this building has a lovely sliding glass door on the second landing, but something is missing! If someone accidentally decided to walk out that slider, he/she would be in for a big surprise. If you have incomplete projects that pose a hazard, be sure to either finish them or make some sort of barrier protection for anyone that might try to use it, such as in the case of the missing platform.

Last but not least, don't forget to use good ladder sense when working on ladders. Refer back to Chapter 4 for some good tips on ladder safety. Improper use of ladders can be very hazardous not only to you but to anyone near you. Before you climb any ladder (Fig. 8-30) be sure that it is positioned correctly and is on firm ground. A few minutes extra in precautionary checking can make the difference between a job well done or a few days in the hospital.

Fig. 8-26. Loose connections for posts or railings require immediate attention and repairs.

Fig. 8-27. Missing portions of stair rails should be replaced or secured to prevent anyone (particularly a child) from falling through.

Fig. 8-28. Steps of uneven heights can be a tripping or falling hazard. Be sure that handrails on each side of the steps are accessible. Find out from a contractor if they can be corrected or improved.

Fig. 8-29. Any sliding door, opening, etc., that opens out *must* have a platform or safe stairways. Make sure that any unfinished projects are safely secured to prevent an unwarranted accident.

Fig. 8-30. Use good ladder sense when working with ladders.

SUMMARY

Now that you have read this chapter and learned about the many hazards that might be in your building (probably more than you had bargained for), you might start to look at your residence or investment property with a sense of foreboding and some apprehension. Please don't! All buildings have defects and deficiencies. At least now you are much better able to cope with any problems that come your way. Be sure to call or write for the many useful, free, or inexpensive publications listed in this chapter. These publications contain more detailed information that will be beneficial both to you and your building.

Maintenance

In Chapter 1 you learned the key step that is first in your maintenance program—namely how to inspect defensively to protect and perhaps even increase the value of your property. In this chapter you will learn how to maintain your building in good health. So that the content is at your fingertips, all building components that need your attention and care have been organized alphabetically, and additional information is provided in appropriate corresponding tables.

Remember, it is important to always follow the manufacturer's recommendations and directions when you work on equipment. Also, because the risk of personal injury is always present, disconnect power cords, shut down systems, pull fuses, ventilate rooms, and generally use common sense when working. Under no circumstances should you put yourself in jeopardy. If you find that a maintenance job is too difficult, by all means never hesitate to call in a professional to do the job for you. Chapter 10 explains what you need to do to get expert advice and help.

Air Conditioners

Clean and replace filters at the beginning of the cooling season and as often as every month during the season. On individual window air conditioners, clean the condenser coil fins with a vacuum cleaner. On a central air system, consult the dealer, maintenance manual, or your serviceman as to specific maintenance steps. See Table 9-1 for some heating and cooling tips. Big savings can be had if you set your temperatures at 78°F for the summer and 68°F for the winter. Once a year check exterior portions for rust spots and sand, prime, and paint those that you find.

Table 9-1. Guide for Heating and Cooling.

✔ Reduce heating costs by turning down your thermostat at night to 55° F.
✔ Use your fireplace in the spring and fall to take the chill out of the house.
✔ Use a glass screen in front of the fireplace to cut heat losses.
✔ Wear warm clothing in the winter to avoid pushing up the thermostat.
✔ Use nature to help heat and cool your building.
✔ Whenever you are out for any length of time, turn the thermostat down.
✔ Make sure that your thermostat is far enough away from windows and doors.
✔ Don't block heating and cooling registers with furniture or drapes.
✔ Maintain and service your heating and cooling systems.
✔ Insulate and weatherstrip your building.
✔ Add storm windows and doors to your building.
✔ Plant trees and shrubs to cool in the summer and break the force of the wind in the winter.
✔ Make sure that your fireplace damper closes tightly.
✔ Caulk and seal cracks and open joints.
✔ Turn off lights that are not being used.
✔ Use exhaust fans to vent excess moisture and heat outdoors.
✔ If necessary, use an attic fan to vent out built-up moisture in attics.
✔ Use cross-ventilation for a natural draft to ventilate your building.
✔ Consider light-colored roofing when your roof has to be replaced.
✔ Think about installing awnings for shade on southern and western exposures.
✔ Insulate heating and cooling ducts.
✔ When you buy appliances, look for approved energy-efficient models.
✔ Clean or change air filters every month during the heating-cooling seasons.
✔ Make sure that you have your heating and cooling systems professionally serviced on an annual basis.

Appliances

As you can see from Table 9-2, appliances, like all building components, have an anticipated useful life expectancy. If you know the life expectancy of each of your appliances, you will be in a much better position to judge whether or not they are worth repairing or whether it is time to put them out with the trash. Most experts agree that repairs to small appliances such as hair dryers, toasters, and garbage disposals should cost less than one-third of the original cost of each appliance. If the cost is one-third or more, buy new ones. For larger appliances such as an electric range, refrigerator, or a dishwasher, the same holds true except the repair costs should be less than one-half of the original cost.

If you are interested in repairing appliances yourself, you can buy manuals from GE and Hotpoint by calling this toll-free number: 1-800-447-4700. Because these manuals are written in such a way that step-by-step procedures apply essentially to all brands, you don't have to worry about which manufacturer made your appliance. If you decide

Table 9-2. Life Expectancies.

Material	Estimated Useful Life (years)
Roofs	
asphalt shingles	15-25
wood shingles or shakes	30-40
tile/terra cotta	40-60
slate	50-80
metal	60-plus
roll roofing	7-10
composition/built-up	10-25
asbestos-cement	40-60
single-ply rubberoid	unknown at this date
Chimneys	
masonry with clay liner	60-100
masonry without a liner	20-30
metal	10-20
pointing up	10-20
Gutters	
seamless aluminum	15-25
galvanized steel	20-30
copper	40-plus
wood	20-30
vinyl	15-20
Siding	
wood	30-40
brick	life of building
aluminum	unknown
vinyl	unknown
stucco	30-40
Decks/Porches	
untreated wood	5-7
treated wood	30-plus
Shutters	
wood	20
metal	20-30
plastic/vinyl	lifetime
aluminum	lifetime

Table 9-2. Continued.

Material	Estimated Useful Life (years)
Windows and Doors	
wood, double-hung windows	40-50
window glazing	5-6
steel casement windows	30-40
jalousie windows	30-40
screen doors	5-8
storm doors	10-15
interior doors (luan)	10
garage doors	20-30
Exterior painting	
wood	5-7
brick	5
aluminum	10-12
Exterior staining	
wood	7-10
Driveways	
asphalt	10-15
concrete	20-30
gravel	10-20
Swimming pools	
pool shell	15-25
pool filter	3-5
pool heater	4-7
pool liner	8-12
Other exterior items	
sprinkler system	15-25
wood fence (untreated)	7-12
pressure-treated fence	30-plus
concrete patio	15-25
tennis court	20-40
Heating and Cooling	
furnaces	15-25
steel boilers	15-25
cast-iron boilers	20-40
heat pump compressors	8-12

Table 9-2. Continued.

electric baseboard	15-plus
wall convection heater	15-plus
air conditioner compressor	10-15
window air conditioner	10-20
air-to-air heat exchanger	15-plus
electronic air cleaner	10-15
expansion tank	30-plus
oil tank	40
buried oil tank	15-25
thermostat	20-plus
combustion chamber	15-plus
stack relay	10-20
flame-retention burner	20-plus
circulator pump	10-20
condensate pump	5-10
humidifier	2-10
heat exchanger	5-30
galvanized heat ducts	50-70
fiberglass heat ducts	40-60

Plumbing

gas water heater	6-12
electric water heater	10-15
oil water heater	10-12
galvanized steel pipes	30-40
copper pipes	50-60
brass pipes	35-45
plastic pipes	35-50
cast-iron drains	50-plus
plastic drains	50-60
septic leaching fields	18-22
submersible well pump	10
pressure tank	20-plus

Electrical

wiring (copper and aluminum)	life of the house
circuit breaker panel	30-40
circuit breakers	25-35
electric fixtures	20-30
doorbell and chimes	7-12
fluorescent bulbs	3-5

Appliances

stoves	15-25
exhaust fan	10

Table 9-2. Continued.

Material	Estimated Useful Life (years)
microwave ovens	10-15
refrigerators	8-16
disposals	5-12
trash compactors	5-10
dishwashers	7-12
dryers	7-12
washing machine	8-12
instant hot water	5-10
dehumidifiers	5-7
humidifiers	5-7
Floors	
oak	life of building
pine	life of building
slate, quarry, flagstone	4-50
resilient (vinyl)	10-15
terrazzo	life of building
carpeting	several years

that a professional should handle the job, get in touch with a factory-authorized repair center. Table 9-3 lists some of the major appliance and equipment manufacturers.

Aluminum Siding

Many communities require that metal-clad buildings be properly bonded with a grounding connection that is attached to the electrical and plumbing system. If you have a building that has metal siding, such as aluminum, and you find that there is no such grounding component, then by all means check with your local electrician or with the city or town hall to see if this is required. Even if it is not mandated, it is a good idea to have it done as there is always the potential for shock hazards from lightning.

Asbestos

Because free-floating asbestos fibers in the air from deteriorating asbestos insulation could contaminate you or anyone in close proximity to it, contact an asbestos abatement company. They can tell you what you can do to prevent further damage to your insulation and additional contamination to your building. If you do have asbestos insulation in your building, it might be wise to have an industrial hygienist test the air to determine the amount and extent of contamination that could exist. Under no circumstances should you attempt to correct the problem yourself!

Attics

Although attics are quite often "out of sight—out of mind," next to the basement it is probably one of the most crucial areas in your building, one that requires inspections,

**Table 9-3. Manufacturers of
Major Appliances and Equipment.**

AMANA
Main Street
Amana, LA 52204
(319) 622-5511

BISSEL
2345 Walker Rd., N.W.
Grand Rapids, MI 49504
(616) 453-4451

BRAUN
The Schwabel Corp.
281 Albany St.
Cambridge, MA 02139
(617) 492-2100

CUISINART
411 W. Putnam Ave.
Greenwich, CT 06830
(203) 622-4608

EUREKA
1201 E. Bell St.
Bloomington, IL 61701
(309) 828-2367

FARBERWARE
Kidd Company
1500 Bassett Ave.
Bronx, NY 10461
(212) 863-8000

FRIGIDAIRE
General Motors Co.
P.O. Box 4900
Dayton, OH 45449
(513) 297-3400

GENERAL ELECTRIC
3135 Easton Turnpike
Fairfield, CT 06431
(203) 373-2211

GIBSON
515 W. Gibson Dr.
Greenville, MI 48838
(616) 754-5621

HAMILTON-BEACH
P. O. Box 1158
Washington, NC 27889
(919) 946-6401

HOOVER
101 E. Maple St.
No. Canton, OH 44720
(216) 499-9200

KIRBY
1920 W. 114th St.
Cleveland, OH 44102
(216) 228-2400

KRUPS
Robert Krups
No. America
7 Pearl Ct.
Allendale, NJ 07401
(201) 863-8000

MAYTAG
403 W. 4th St., No.
Newton, LA 50208
(515) 792-7000

MR. COFFEE
North America Systems Inc.
Bedford Heights, OH 44146
(216) 464-4000

NORELCO
North America Phillips Corp.
High Ridge Park
Stamford, CT 06904
(203) 329-5700

NORGE
Herrin, IL 62948
(618) 988-8431

OSTER
5055 N. Lydell Ave.
Milwaukee, WI 53217
(414) 332-8300

Table 9-3. Continued.

PANASONIC
1 Panasonic Way
Secaucus, NJ 07094
(201) 348-7000

PROCTOR-SILEX
1016 T W 9th Ave.
King of Prussia, PA 19046
(215) 265-8000

QUASAR
9401 W. Grand Ave.
Franklin Park, IL 60131
(312) 451-1200

ROPER
1905 W. Court St.
Kankakee, IL 60901
(815) 927-6000

SPEED QUEEN
Shepard & Hall Sts.
Ripon, WI 54971
(414) 748-3121

SUNBEAM
5400 W. Roosevelt Rd.
Chicago, IL 60650
(312) 854-3500

TAPPAN
Tappan Park
Mansfield, OH 44901
(419) 529-4411

TOSHIBA
82 Totowa Rd.
Wayne, NJ 07470
(201) 628-8000

WARING
Route 44
Hartford, CT 06057
(203) 379-0731

WESTBEND CO.
West Bend, WI 53095
(414) 334-6922

WESTINGHOUSE
Gateway Center
Westinghouse Bldg.
Pittsburgh, PA 15222
(413) 255-3800

WHIRLPOOL
Administrative Center
Benton Harbor, MI 49022
(616) 926-5000

appropriate maintenance, and repair work. During or after a heavy rainstorm, check the attic for leaks. While you are checking for leaks, be alert for signs of decay in framing members, deterioration in chimney areas, and for visible skylight from open flashing joints. If you spot any areas of concern, mark them down for future repairs, caulking, and sealing. While on an inspection tour of the attic, check that you have sufficient insulation in place, including the access panel. If you notice that you have minimal or no ventilation, call in a carpenter to add more vents to air out any trapped moisture that might be lurking in the attic.

Basements

During your routine inspection tour, watch carefully to make sure that you are not storing flammable materials near a heat source, such as your boiler. Try not to keep and store everything that you have ever owned or found. This stored clutter could be the catalyst for a future fire. In the basement, always be on the lookout for any sags or gaps between beams and girders. If you see any areas of concern with structural members, have a contractor come in to see if you need additional support posts. Always check openings such as windows and bulkheads for air leaks and make sure that they are well sealed and weatherstripped to keep out the weather. If you have insulation in the basement, remember that the vapor barrier on the insulation should always face the heated sides of the building. Musty odors in your basement can be telling you that a dehumidifier is needed. After or during a good rainfall, check the basement for signs of water entry. If you do have some water coming into your basement, be sure to check back in Chapter 2 for possible remedies.

Bathrooms

Caulk and seal open joints between tiles and fixtures with a good quality caulking compound such as silicone. Grout tile joints as they become loose and open. Missing caulking and deteriorated grout are the major source of water damage to walls and ceilings. A handy tip for caulking and sealing tub-tile joints is to apply the caulk while the tub is full of water—preferably with you in the tub—so the tub is at its lowest point. After the caulk has set, drain the tub. Any further movement will not adversely affect the caulking because the seal was applied while the gap was at its widest.

Use a discarded toothbrush to clean dirty grout joints. You can buy a specially formulated grout cleaner at most hardware stores. Remember, never use abrasive cleansers on fixtures; you will scratch and mar the finish.

Baking Soda

Common household baking soda is a nontoxic, versatile cleaner and absorbent. It can be used as a substitute scouring powder on most hard surfaces. Use it dry—it will not scratch. Put it into the refrigerator to absorb odors and pour it down your drains in place of more caustic drain cleaners. Try pouring a handful or so down your drains once a week, washing it down with hot water. Be sure to have some in your building; if you have not used it before, give it a try—you might like it.

Blowers

Blowers in furnaces require maintenance at least twice a year. If your heating system is a forced warm-air system, follow these suggestions. Every six months or so, lubricate oil cups with a few drops of light oil (do not overoil). Vacuum air slots and fan blades to remove accumulated dust and dirt. Periodically, check the fan belt for frayed or worn spots and replace the belt if it is worn out. Test its tension by placing your finger against it. It should not give more than ½ inch, and if it does, have your serviceman tighten it.

Boilers

Refer to your boiler maintenance manual or the heating system maintenance suggestions in Chapter 7 to see how to keep your boiler in good working order. Both boilers and burners should be inspected and serviced by a qualified serviceman at least once a year.

Brick

To clean brick, scrub it with a solution of water and washing soda or trisodium phosphate (TSP). Oily stains can be removed by applying a paste of TSP and water, and soot and smoke can be removed with a paste consisting of ammonia and powdered pumice. To get rid of efflorescence (a white powdery substance found on masonry surfaces), use a mixture of one part muriatic acid and nine parts water. Because muriatic acid is toxic, it might be a good idea to have a professional handle this chore for you.

Burners

Whereas oil burners need to be serviced every year, gas burners should do just fine with service every two or three years. It is best to leave that kind of job to the professional, though, who has the necessary expertise and the right tools to work with. When the burner is being tuned up, have your serviceman check the efficiency of the entire system. He should then use the test results to make the necessary adjustments to fine-tune your burner.

Cabinets and Countertops

Don't use abrasives to clean cabinets and countertops. For plastic-coated units, use a detergent solution. Apply paste waxes containing either carnauba wax, beeswax, or paraffin and buff to bring out the shine in the wood. If you find gouges in wood surfaces, use a wood putty that matches the wood of the cabinet. When the putty is dry, sand down to a smooth surface, then refinish and rewax. If the Formica top has loosened up, apply a good quality bonding glue to the loose section and clamp it down. Follow the directions on the can for long-lasting results.

Circuit Breakers

Trip circuit breakers off and on several times every 6 months to "exercise" them and to make sure that they are functional. Circuit breakers that are faulty in that they trip off by themselves, are difficult to trip, or have substantial corrosion on them should be replaced as soon as possible.

Chimneys

Every year inspect your chimney for soot and creosote buildup and have the flues cleaned by a professional chimney sweep, if necessary. Check for loose bricks, deteriorated joints, damaged brick and cap areas, and open flashing. Never install an antenna on a chimney or vent pipe. If you have such a partnership at the moment, dissolve it before you damage the chimney and open up flashing joints. Make it a habit to keep up with these annual maintenance chores and repairs.

Chlorine Bleach

Liquid chlorine bleach mixed with water is a very effective treatment to get rid of mildew from either the exterior or interior of your building. Try the following solution not only to get rid of mildew but also as a potent cleaning solution for all kinds of surfaces: one quart liquid bleach, three quarts water, and a half a cup of TSP or Spic and Span. Because chlorine bleach is toxic, keep it away from children and pets. Also, avoid contact with skin and eyes and make sure that you avoid breathing in chlorine vapors.

Condensation

Moisture trapped in your building will most likely condense and thus cause water problems. Be sure to have sufficient ventilation not only in your attic, but also in any crawl spaces. Opening a window now and then for very short periods of time during the winter can also help to vent out trapped moisture. Don't forget that the judicious use of dehumidifiers is another great way to cut down on moisture in a building. Review Chapter 3 for a detailed account on condensation.

Crawl Spaces

If your building has a crawl space, ventilate, ventilate, and again, *ventilate*. This is one of the key things that you must do to keep your home healthy. Be sure that all crawl spaces have plenty of openings in them to allow moisture to escape. In addition, place a polyvinyl plastic sheet over exposed soil to keep down the amount of moisture in the crawl space. Lap the sheets several inches and hold them down with bricks or other solid objects. Also, inspect on an annual basis for signs of decay, wood-boring insect activity, and damage.

Decks

Inspect and paint or stain untreated wood decks or porches annually. For repairs, or if you have to replace some sections, only use pressure-treated wood, which is particularly important if the wood is in direct contact with the soil. In fact, when you buy the wood, tell the dealer that it will be in direct contact with the soil. Some forms of pressure-treated wood are not meant for soil contact. Also, be sure that safety handrails are secure and that spindles are properly spaced.

Dehumidifiers

Dehumidifiers are very important units that remove moisture from a building. At least twice a year, remove the cover from the unit and clean the coils. Check the drain hole and if it is plugged, clean it out. Lubricate all oil port holes (see the motor shaft on the unit) with a few drops of 20-weight oil. As is true with any appliance, follow the manufacturer's recommendations and directions for efficient operation.

Dishwashers

Weekly maintenance of dishwashers includes getting rid of clinging food particles, seeds, and the like on the spray area and on the filter screen. Remember that a dishwasher can't operate efficiently with a clogged-up filter. If you lower the temperature on the water heater to 120°F, you will save money on your energy bill. If you find that your dishes aren't coming out as clean as you would like them to be, however, increase the temperature to 130°F. To unclog your dishwasher drain, sprinkle one cup baking soda near the drain and pour a kettle of boiling water over it. Never use commercial drain cleaners because they will corrode the aluminum and rubber parts.

Garbage Disposals

Always run your disposal with cold water so that you can get rid of grease. Grease needs cold water to congeal. Here also, never use commercial drain products. Not only will they damage the working parts, but there is always the risk of some splashing back into your face. Instead, grind up lemon or orange peels to clean and sweeten the interior of the disposal. You can also clean it by grinding up ice cubes sprinkled with scouring powder. In the event that your disposal jams up, turn it off and let it cool down. Then press the reverse button if there is one (look underneath the disposal) or use the end of a broom handle to force the blades to reverse. That should do the trick.

Doors

Doors undoubtedly get battered about more than any other building component. They are forever being opened and closed and, yes, sometimes even kicked. Every year check the hinges to make sure that they are aligned correctly and secured tightly to the frame. If doors stick, take some sandpaper or use a plane to reduce the edge that is catching. Be sure to repaint or stain this area so that it can't absorb moisture, which would eventually cause the door to swell and warp. Also make sure that the top and bottom of the door fit snugly into their frames. This is particularly important for exterior doors because they help keep out unwanted weather conditions. Each spring and fall inspect their weatherstripping for wear and tear. Tighten up the metal kind and prolong the vinyl-rubber type with an occasional application of petroleum jelly.

Downspouts

Make sure the water coming from your gutters into your downspouts doesn't seep into your basement. Check that the ends of the downspouts extend far enough away from the building—at least several feet. Splash blocks or, better yet, 4-foot extensions connected to the base of the downspout bases can divert water away from your foundation. If your downspouts lead into dry wells (buried holes in the ground), you might have the makings of foundation seepage problems. Quite often, over a period of time, dry wells plug up with sediment so that water backs up into the basement. Should you have trouble with such backups, simply cut off the base of the downspouts a few feet above the dry well connection and attach 4-foot extensions to them, which then would carry water safely away from your building.

Drains (Plumbing)

Don't pour any liquids or even solids down your drains that could harm the plumbing system. That goes for grease, paint, food, pesticides, lubricants, oils, and so on. Try to keep drains free from hair (yours and your pet's) as well. To prevent a buildup of grease or soap, pour 3 tablespoons of baking soda mixed with hot water into each of your drains and let them sit for 20 minutes. Flush this solution with hot water after some time has elapsed. Do the procedure every month. Never use lye products or lye derivatives, particularly with plastic plumbing drains. Other safe products to use to avoid clogged drains are: washing soda, vinegar, and/or ammonia, each mixed with boiling water. Sometimes weekly applications might be needed. Be sure to use drain strainers to catch unwanted items before they bid farewell down the drain. If you have no drain strainer, pay your local hardware store a visit for replacements.

Driveways

Driveways, by their very nature, will crack, sink, heave, and get damaged in all sorts of ways. Besides always being exposed to the outside elements, they also suffer from the heavy traffic and the natural movements in the earth below them. You can easily maintain your asphalt driveway by following the simple instructions given here. A good quality blacktop sealer is the first line of defense against the weather, and it should be applied every 3 to 4 years or as often as needed. Small cracks in the asphalt can be repaired by first cleaning out all the loose matter in the cracks, filling them with a butyl cement, and smoothing the top down with a putty knife. Larger cracks can be filled with blacktop patching compound by packing it down with a trowel. Lay a board over the patch and drive your car over it several times to really compact it. Driveways made of concrete can be repaired by using a ready-mix cement that you can buy at any lumberyard. Be sure to follow the directions on the package.

Dryers (Clothes)

Your dryer is an appliance that is relatively simple to maintain. Once a year remove the exhaust hose and clean out all accumulated lint and dust. Make sure that your electric or gas dryer vents directly to the outside and not into the building. (One of the main sources of indoor humidity is from improperly vented dryers.) Because accumulated lint decreases the machine's efficiency and is also a potential fire hazard, clean the lint screen on a weekly basis or, better still, after each use. If you find that your clothes tear, inspect for jagged edges on the interior of the machine. Sometimes objects lodge in the rim, and if they are left there, they can tear up your clothing as the unit spins. Avoid using antistatic sprays in your dryer—they tend to clog up the filter.

Dry Rot

Dry rot is actually a misnomer because wood will not rot if it is dry. Only with repeated wettings and drying out will it start to decay. Untreated wood decks, porches, siding, and trim must be kept painted or stained to avoid water damage. Decks and porches (unless they are constructed with pressure-treated wood) need annual attention.

Efflorescence

If you find a white, powdery substance on your foundation walls and your basement floors, you are probably dealing with efflorescence, which is a deposit of salt that remained after the water had dried. The key step is to correct the cause of the water getting through your foundation in the first place (see Chapter 2 on wet basements). Once that is done, use a solution of 1 part vinegar to 3 parts water for insignificant amounts of efflorescence and a solution of 1 part muriatic acid to 9 parts water for substantial amounts. As noted earlier in this chapter, muriatic acid is potentially harmful to use; you might wish to have a professional do the work.

Electric Heat

Electric heat is very clean and relatively efficient as long as the building is sufficiently insulated. In many parts of the country, however, it is rather expensive to heat with.

Individual room thermostats can isolate heat to only areas you want it. If your system does not have a thermostat in each room, consider having an electrician zone it for you. To maintain electric heat, vacuum baseboards monthly and keep unused room thermostats off or at a low setting.

Electrical System

Every year check the grounding connection to your waterline or to the driven rod in the soil adjacent your property. Make sure that it is tight and still properly secured. Trip all circuit breakers twice a year to "exercise" them. On a monthly basis, trip and test the ground fault circuit breaker. If you blow fuses all the time, have a licensed electrician troubleshoot the circuits for you. Any time wall outlets or switches give off odd odors, unusual sounds, or feel unusually warm to the touch, you have cause for alarm. Don't hesitate to call in your electrician to find out the cause. If you need additional outlets, don't depend on flimsy lamp cord additions—have proper wiring installed by a professional.

Entrances

All entrances and exit stairways should have safety handrails. Make sure that the rails and spindles are securely attached to the stairs and to the building itself. Repair steps and platforms that are damaged as soon as possible; if walkways are heaved and buckled, get them in ship-shape condition so no one will take a tumble. If during your inspection tour you note that some of your steps are 5 inches in height, for example, while others are 6 or 7 inches, you might want to correct this uneven condition before this inconsistency causes an injury.

Expansion Tanks

Expansion tanks are part of forced hot-water heating systems. Sometimes the tank loses its air cushion, which provides a buffer for the system when it is under operating pressure. If you find that your expansion tank is full of water rather than half full of water and air (you can tell sometimes when the relief valve starts dripping water), call your serviceman in to correct the situation.

Exterior (Building)

Table 9-4 recaps the major maintenance jobs for the exterior of your building. Remember that conditions on the outside of your building reflect on the inside. The first step in your defense is to make the exterior tight to keep the heat in and the weather out. Check Tables 9-5 and 9-6 that are based on the months and on the seasons respectively. Review all of these tables and make sure that you keep up with required maintenance chores.

Fans

Fans are very handy to have when you want an extra cool breeze. So that they will work properly when you need them, clean the blades by unplugging the unit, removing the blades, and washing them in hot, soapy water. If you can't remove them, wash the blades in place. Oil the fan according to the manufacturer's instructions and lubricate accessible bearings.

Table 9-4. Exterior Maintenance Chart.

✓ Inspect roof shingles twice a year (before and after the winter).
✓ Repair and replace damaged roofing.
✓ Before each heating season, have chimney cleaned and inspected.
✓ Repair defective flashing on chimney and roof areas.
✓ Have necessary masonry repairs done to foundation and chimney.
✓ Use caulking compound to seal open joints and gaps.
✓ Repair defective screens and cracked window panes.
✓ Lubricate doors and windows (hardware).
✓ Trim back all overgrown vegetation near building.
✓ Regrade all foundation soil to allow a slope away from the building.
✓ Provide downspout extensions to carry water away from the foundation.
✓ Clean, service, and repair gutters.
✓ Repair decaying wood siding and trim.
✓ Provide several inches of space between low wood areas and soil.
✓ Add ventilation to unvented crawl space and attic areas.
✓ Sand, prime, and paint areas that show peeling paint.
✓ Have well water tested annually by a testing company.
✓ Inspect private sewer systems annually.
✓ Have private sewer systems pumped out every three years.
✓ Provide weep holes in storm windows.
✓ Remove antennas from chimney and vent pipes.
✓ Remove debris from foundation window wells.
✓ Provide covers for foundation window wells.
✓ Check for wood-boring insect activity and damage.
✓ Continue your preventive maintenance program for insects.
✓ Keep tree limbs away from roofs and building areas.
✓ Paint or stain exterior exposed wood decks and porches annually.
✓ Weatherstrip doors and windows.
✓ Repair or replace defective fencing.
✓ Patch damaged driveways and walkways.
✓ Make entrances safe to use (repair steps, add railings).
✓ During the off-season, cover air conditioners to prevent rusting.
✓ Drain outdoor waterlines before winter.
✓ Be sure that your sump pump discharges several feet from your foundation.
✓ Wash mildew off siding and trim.
✓ Fertilize plantings as required.
✓ Provide ground coverings and plantings to control erosion.
✓ Keep several inches of clearance between low wood and soil.
✓ Use pressure-treated wood for exterior projects.

Table 9-5. Maintenance Calendar.

January

- ✓ Monitor heating system.
- ✓ Change filters on warm-air systems.
- ✓ Make sure pumps and motors are well lubricated.
- ✓ Blow off low-water cutoff valve on steam systems as required.
- ✓ Check for drafts and air leaks around doors and windows.

February

- ✓ Recaulk open or damaged joints if necessary.
- ✓ Have flues cleaned if using a solid fuel stove.
- ✓ Keep walkways free of ice and snow.
- ✓ Get some of your inside painting done.
- ✓ Recaulk and grout bathroom tiles.

March

- ✓ Tune up lawn mower.
- ✓ Sharpen garden tools.
- ✓ Get some of your inside carpentry work done.
- ✓ Inspect and adjust window hardware.
- ✓ Start or complete projects such as window boxes for spring.

April

- ✓ Have central air conditioning serviced.
- ✓ Clean and make service adjustments to window air conditioners.
- ✓ Clean and service gutters. Oil wood gutters with linseed oil.
- ✓ Provide extensions for downspouts.
- ✓ Check for winter damage and make repairs.

May

- ✓ Wash exterior siding and trim with liquid bleach and water to remove mildew and dirt.
- ✓ Seal cracks in asphalt driveways and walks with a sealer.
- ✓ Stain or paint decks and porches.
- ✓ Repair screens and replace cracked window panes.
- ✓ Sand, prime, and paint exterior metal railings and trimwork.

June

- ✓ Trim back vegetation next to foundation and siding.
- ✓ Be on the lookout for wood-boring insects.
- ✓ Replace decayed wood with pressure-treated wood.
- ✓ Start exterior carpentry jobs, if not already begun.
- ✓ Rescreen damaged screening on vents.

Table 9-5. Continued.

July

✓ Double-check the operation of garage doors.
✓ Weatherstrip windows and doors, if needed.
✓ Continue with your pest control program.
✓ Inspect and service security systems, locks, and alarms.
✓ Clean dehumidifiers.

August

✓ Continue harvesting crops.
✓ Think about adding insulation to areas that need it.
✓ Have cesspool or septic tank pumped and inspected.
✓ Have masonry repairs done to chimney and fireplace.
✓ Regrade foundation soil to keep basement dry.

September

✓ Scrape, prime, and paint exterior siding and trim.
✓ Start putting your landscaping to bed.
✓ Think about winterization (caulking and weatherstripping).
✓ Clean windows.
✓ Lubricate windows and doors.

October

✓ Start to take in patio furniture.
✓ Close up pool for the season, if still open.
✓ Think about insulating heat pipes, water pipes, and ductwork.
✓ Add a rust inhibiter to the oil tank to prevent internal rusting.
✓ Repair fencing.

November

✓ Continue winterization program.
✓ Shut off all outside faucets and leave spigots open.
✓ Get snowblower in tip-top shape for the winter snows.
✓ Continue with necessary pruning and trimming of trees and shrubs.
✓ If not already done, get heating system tuned up.

December

✓ Make sure attic access door is insulated.
✓ Service humidifier(s).
✓ Clean stove vent hood screen.
✓ Have solid fuel stove cleaned and inspected.
✓ Cover window air-conditioning units with covers.

Table 9-6. Four-Season Maintenance Chart.

Fall

✓ Clean and service gutters.
✓ Clean out foundation window wells.
✓ Remove dead limbs from trees.
✓ Trim back all overgrown vegetation near buildings.
✓ Make repairs to window putty, caulking, and paint.
✓ Service and/or install storm windows.
✓ Shut off outside hose bibbs prior winter.
✓ If not already done, have heating system tuned up.
✓ Winterize and tighten up the exterior of the building.
✓ Caulk and seal any open joints or cracks on the exterior.
✓ Inspect chimney for loose or damaged bricks.
✓ Check for blocked flues in the chimney.
✓ Prime and paint areas with peeling paint.
✓ Check roof for potential winter leaks.
✓ Make sure antennas are secure.

Winter

✓ Paint interior areas that need it.
✓ Get tools in shape for the coming spring and summer.
✓ Check lamp cords, plugs, and sockets for defects.
✓ Double-check outlets for proper grounding.
✓ Test and operate sump pump.
✓ Double-check to make sure all exterior faucets have been drained.
✓ Make sure that you have sufficient insulation in the attic.
✓ Add insulation where needed.
✓ Check windows for drafts and repair as needed.
✓ Check tub and shower caulking and grout and reseal if required.
✓ Vacuum refrigerator coils.
✓ Monitor heating plant weekly during the heating season.
✓ Make house tight to keep out uninvited guests (mice).
✓ Keep sidewalks and walkways free of snow and ice.
✓ Double-check pipes and drains in exposed areas for possible freeze-ups.
✓ Use sand instead of corrosive rock salt for slippery areas.

Spring

✓ Have septic/cesspool pumped out every third year.
✓ Inspect septic/cesspool tank annually.
✓ Double-check roof drainage for winter damage.
✓ Be on the lookout for wood-boring insect activity.
✓ Prune damaged trees, shrubs, and plants.
✓ Check for winter damage to roof, chimney, and building in general.

Table 9-6. Continued.

✓ Have cooling system serviced.
✓ Clean individual window air-conditioning units.
✓ Check attic for evidence of leaks.
✓ Inspect for flashing damage, repair if necessary.
✓ Oil and lubricate motors and pumps.
✓ Check smoke detectors according to manufacturer's directions.
✓ Rake up winter debris from yard.
✓ Regrade foundation soils to keep water out of basement.
✓ Make repairs to walkways, driveways, and patio areas.
✓ Double-check building caulking; recaulk if required.

Summer

✓ Start planning for fall painting chores.
✓ Reseal asphalt driveway and patch concrete walks.
✓ Fill in all sinkholes in yard.
✓ Check for rot and insect damage.
✓ Keep all low wood away from soil.
✓ Repair and replace decaying wood with pressure-treated wood.
✓ Have chimney cleaned and flues inspected by chimney sweep.
✓ Keep vegetation away from the building.
✓ Repair or rebuild fences.
✓ Continue maintenance on swimming pool.
✓ Prune trees and shrubs as required.
✓ Stain or paint decks and porches.
✓ Remove mildew and mold from building siding and trim.
✓ If required, have building treated by an exterminator.

Fences

Fences require annual inspections and maintenance repair work. Paint or stain wood fences yearly—unless they are made of pressure-treated wood. Replace decayed or damaged wood sections only with pressure-treated wood. Check metal fences annually for rusting sections. If you find rust, sand down to bare metal, prime with a good metal primer, and then apply two coats of an appropriate metal paint.

Fire Extinguishers

Every month or so check the indicator on the pressure gauges to be sure that your extinguisher is properly charged. Make sure the lock pin is firmly in place. Look over the nozzle to see if it is clogged. By no means test the extinguisher itself because you will lose a substantial amount of pressure. Once every 6 months have the extinguisher refilled and charged by a service company.

Fireplace

Use common sense when you light your fireplace. Never burn trash, green wood, pressure-treated wood, or anything that will cause excessive sparks or creosote buildup.

Treated wood, for example, has chemicals that could seriously harm you if you inhaled the fumes. Before you light up the first fire of the season, ensure that the fireplace is in good working condition. Check the brick, the mortar joints, and the walls once more to make sure you have not overlooked any defects. The damper should operate properly because a loose one could surely cost you a lot of energy dollars. Oil the damper rod yearly to make it work smoothly. Keep the flue liners free of soot and creosote, and it is not a bad idea to have a professional chimney sweep clean all of your flues on an annual basis or as often as they need sweeping. You can help out by tossing a handful of salt on the fire once in awhile. Don't do this if you have a metal chimney flue as it will corrode it. Always remove the built-up ashes in the cleanout at the base of the chimney. Add them to your garden soil and you just might win a prize for the largest tomato on your street.

Flashing

All joints, connections, and intersections between parts of your building should have flashing to seal out water. Areas such as chimneys, vent pipes, skylights, door and window openings, and places where two different materials meet, such as brick veneer against wood siding, should be properly flashed. Leaks into your building are very likely to occur at these points, one reason why it is important to annually check and repair deterioration in these areas. Sometimes the repair work can be as easy as just tapping down loose flashing, while other times it might require a professional replace it. One quick and easy temporary solution is to apply roofing cement to damaged areas. This usually works for awhile as long as you remember to do it every year. Eventually, though, you must replace the flashing. When replacing or repairing flashing, be very careful because most of these areas are relatively high. You might want to hire a professional to do this hazardous work.

Floors

Table 9-7 lists the more common types of floor coverings and provides a guide for floor care. Always refer to your dealer's recommendations for cleaning and waxing.

Foundations

Foundations will crack as they settle in the soil and because of water pressure. Most cracks are considered minor, and you can easily fill them with a patching cement. Cracks that are over ¼ inch in size and appear to be getting larger should be checked out by a professional who can determine the extent of the problem. Seal gaps around openings with silicone caulking or weatherstripping. To avoid future damage, keep all roof drainage runoff away from your foundation by grading all soil adjacent to the foundation to slope several feet from the walls and by providing extensions for downspout bases.

Framing

If untreated wood framing gets wet on a regular basis, it is likely to rot and draw wood-boring insects. This holds true whether the wood is on the outside or the inside of your building. On the outside, keep all soil several inches away from low wood members,

Table 9-7. Floor Maintenance.

Type	Care
Asphalt tile	Mop with water and mild detergent. Use only a water-based wax. Alkaline or nonalkaline cleaners are safe.
Brick	Seal with masonry sealer. Wax with water-based wax if floor is sealed.
Carpeting	Vacuum on a regular basis and don't use too much water when shampooing.
Ceramic tile	Mop with water and mild detergent. Polish with a buffing machine or by hand. Regrout joints and apply tile sealer.
Concrete	Seal with masonry sealer. Paint with a masonry paint.
Cork tile	Use only solvent-based waxes and cleaners. Apply wood sealer.
Hardwoods	Apply wood sealer. Use only solvent-based waxes and cleaners.
Rubber tile	Wax frequently with only water-based wax and clean only with water-based cleaners.
Slate, flagstone, and quarry stone	Use a sealer specified for stone floors. Clean with water and a mild detergent. Wax with a water-based wax.
Terrazzo	Apply a sealer. Wax with a water-based or solvent-based wax or polish.
Vinyl	Mop with a water-based polish and wax with a water-based polish.
Vinyl-asbestos	Same as vinyl.

which includes foundation windows, bulkheads, door trim, siding, base of stairs, porch and deck framing, and lattice work around crawl spaces. Insofar as the inside framing is concerned, areas such as the base of partition walls, baseboards, ends of wall panels, wood platforms used to support appliances, and the base of the basement stairs all should be kept dry. Paint or stain these areas and only use treated wood for repairs and replacements. Following these simple steps judiciously will discourage rot and most wood-boring insect activity.

Freezers

A freezer is a big ticket item and should therefore be treated with love and tender care. One of the first things you should check is the temperature of the freezer. A typical noncommercial one should have a temperature of 0°F. You can test it to see that it is indeed 0° by inserting a thermometer between two frozen packages and leaving it there for 24 hours. If you find that it is warmer or colder, adjust the freezer thermostat and retest it again 24 hours later. Continue until you reach the desired temperature.

Every 6 months, wipe the interior walls with a soft sponge or cloth soaked in a solution of baking soda and warm water. Cap this off by wiping again with alcohol or vinegar to keep the freezer's contents from sticking to the damp walls. Don't forget to wash the door gasket with warm water and a mild detergent. Annually, vacuum the condenser and clean the coils with a special brush that is available at most appliance stores. To keep things safe, always unplug your appliance while you are working on it.

Fuses

If you blow fuses in your service panel, be sure to replace them with ones of similar size. Never substitute a 15-amp fuse or a 20-amp fuse with a 30-amp one because it will not blow if there is an overload. The undersized wiring that serves that fuse, however, is likely to heat up and cause a fire. You might be interested in installing Type-S nontamperable fuses that come with adapters that permanently stay in the fuse socket so that fuses of other sizes just won't fit into them. You can buy these at any electrical supply store.

Grading

Grading is crucial in keeping your basement dry. Slope the soil around your foundation away from your building. Watch out for erosion along the foundation. If little ditches develop, standing water could seep into your basement. Every 6 months or so walk around your foundation; if you find low spots, fill them in as soon as possible. Don't be afraid to pack the soil against the foundation wall nice and tight; be sure it gently slopes several inches (better still, a few feet) away from your building.

Graphite

Use powdered graphite to lubricate your locks rather than the liquid kind so as not to attract and hold dirt. Be sure to lubricate two or three times a year, particularly during the winter when locks stick or freeze more readily. Just take the container that the graphite comes in, put its nozzle into the lock opening, and squeeze out a couple of jolts. Your locks should be as good as new after this.

Ground-Fault Circuit Interrupters

A GFCI device is a special safety outlet that is now mandatory on new construction in bathrooms, garages, exterior receptacles, swimming pools, and in the kitchen. What it does is simply to save lives. It monitors the current flow in an outlet or in the service panel. If there is even the slightest discrepancy (a drop in the current), it shuts down all the current in that particular circuit in about a fortieth of a second. Most manufacturers suggest that you test them on a monthly basis to make sure that they are functional. Just push in the test button on the outlet and then reset it. If for some reason one of them does not work, have it replaced immediately.

Grounding

All electrical systems must be grounded to prevent serious shock or death from malfunctioning outlets, receptacles, or equipment. How to check for grounding can be reviewed under "Electrical Systems" in this chapter. If for some reason you find your grounding connection is loose or disconnected, reattach it to the street side of your water meter. Some local electric codes now require even older homes to have a jumper cable as an added protection. This cable will be attached to both the street side and the house side of the water meter. Check with your local electrical/building department to see if they require such a hookup. Wall outlets must also be grounded; to test if they are, use a receptacle analyzer. Outlets that are not grounded, particularly those in your kitchen or bathrooms, should be corrected by an electrician.

Gutters

Check your gutters at least twice a year (spring and fall—more often if your building is surrounded by trees) for piles of leaves and debris. After cleaning them out, flush them with a hose. If you don't have leaf traps over your downspouts, consider adding them. They are inexpensive items that fit into the top of the gutter-downspout connections so that the downspouts won't plug up. See to it that the gutters slope in the direction of the downspouts and that the gutter hangers are in tight and secure to the building. Annual maintenance will consist of spot-painting worn areas, patching holes, and replacing damaged sections. Oil wood gutters at least twice a year with a 50/50 solution of linseed oil and turpentine. Don't be stingy with the amount of solution you put on and allow it to soak into the wood. You can use roof cement to repair all kinds of gutters (metal, wood, and vinyl). Most experts agree that if your building does not have a substantial overhang, a gutter system would be an asset to your building.

Heat Exchanger

Once a year have your serviceman clean the heat exchanger in your heating system. Piled up soot and carbon act as an insulator and decrease the efficiency of heat transfer to your building. Do not waste energy dollars up the chimney and have this crucial area serviced.

Heat Pump

Maintain your heat pump as you would maintain your central air-conditioning system. Before the cooling season, and now and then during the season, clean or replace the filter. Also remove debris that you find around the compressor and keep shrubbery trimmed back to allow free flow of air around the unit. A special word of caution here: Never operate the heating mode during the summer months or the cooling mode during the winter. If you do, you are apt to damage the compressor, which can translate into major costs.

Hot-Water Heating System

Your heating system will not only live longer if you carry out the necessary maintenance jobs but will reduce repair and heating costs as well. Oil the circulation pump twice a year with the type of oil that your serviceman recommends. Bleed each radiator at the beginning of the heating season and again, if necessary, during the season. Check the boiler gauges on the boiler to make sure they do not read high or "danger." If you see the pointer approaching the danger mark on the gauge, shut down the system and call your serviceman. As is true with all heating systems, have yours inspected and serviced annually.

Hot-Water Tank

Hot-water tanks last from 5 to 10 years (even longer than their estimated useful life). You can give your tank extra years of service by draining off a bucket of water every 6 months, more often in areas with a high mineral content in the water. Draining will get rid of built-up sediment in the tank. To drain off the water, place a bucket under the drain valve at the base of the tank and fill it; usually one bucket will suffice. Every 6 months you might also want to check the relief valve on the tank. Lift the lever up and a spurt of water should shoot out of the valve. If after lifting the valve lever nothing comes out, have your plumber come in to determine if the valve needs replacement. If your hot-water tank and hot-water lines are not already insulated, you might want to look into that possibility. You will get your investment back in a very short time from the energy savings.

Humidifiers

Lubricate all oil cups of your room humidifier with a few drops of 20-weight oil every 6 months. Sealed oil compartments, which some humidifiers have, need not be oiled. Lubricate the shaft of each roller of your unit with a dap of petroleum jelly. Also try to keep your humidifier clean by following the manufacturer's recommendations. If you add water softener to the reservoir water, you might be able to minimize mineral deposits and thus prolong the life of moving parts. To add a sweet smell to the appliance, drop some bottled lemon juice into the water in the reservoir cup.

Humidifiers in furnaces require a goodly amount of maintenance and inspection, especially if the water is high in minerals. A humidifier that is mounted in a furnace and that is not maintained could in time destroy the furnace. Water leaking onto the heat exchanger could rust it out in a short time, requiring major repairs or replacements. On an annual basis have your serviceman inspect and service the humidifier. You must keep the unit clean and unclogged by draining and cleaning the water pan on a regular basis. Also periodically work the float arm manually to make sure that it is still functional. Lubricate the motor with the recommended oil and keep the pads and plates clean and free of mineral deposits.

Ice Dams

Make sure that you have sufficient insulation and ventilation in your attic areas to avoid the curse of ice dams. What you are looking for is a free flow of air in the attic. Make sure that insulation has not been stuffed into the overhanging eaves where it would block the incoming air from the soffit vents. In addition, you want cross ventilation that will move air by natural convection in and out of the attic. If you have been cursed with ice dams in the past, check for ample ventilation and sufficient insulation. If you are lacking in either department, have a contractor take a look and see what can be done to alleviate the problem.

Insects

Table 9-8 lists the most common insects that you might come in contact with. Their essential living habits and suggestions as to how to get rid of them will be found in this table. If you decide to use pesticides, follow the directions on the packages as well as those suggested here. Always use common sense, though, and if you have any doubts, contact a professional exterminator.

Guidelines on pesticide usage:

- Before you use any pesticide, try sanitation and mechanical alterations for methods of control—keep all food in proper containers, make sure screens are not damaged, keep wood away from the soil.
- If you decide to use a pesticide, pick one that is formulated for the particular pest that you are trying to control.
- Avoid contact with any pesticide and don't inhale its fumes.
- Keep pesticides away from children, pets, food, and water, as well as cooking utensils and dishes.
- Never smoke while using pesticides; wash your hands after you are through spraying and before you have a cigarette.
- Only use as much pesticide as is recommended on the label.
- Never spray near an open flame, a furnace, or a pilot light.
- Never spray on windy or raining days.
- Do not reuse an insecticide container for other purposes.
- Never use insecticides that have a very long residual life, particularly near where children play, where pets live, or near a garden.
- Change your clothes after spraying and store pesticides in a safe storage area.
- Don't hang up chemically treated strips in rooms that are used all the time by children, the elderly, or where cooking or eating occurs.
- Never stay in a room that has just been treated with a space spray can (bomb). Close the door tightly for at least an hour and then air it out thoroughly.
- Only dispose of pesticides and insecticides and their containers in a way that is recommended by their manufacturers.
- Never flush insecticides down toilets, sewers, drains, or into waterways.
- If you feel ill after you have used an insecticide, contact your doctor immediately.
- Make sure you hire an exterminator that is certified and licensed as well as insured.
- Protect your body from contact with pesticides. If you accidentally make contact, wash off contacted areas immediately with soap and water.
- Always work in a well-ventilated area.
- Be sure to always read the directions on the label as well as following them when using pesticides.

Table 9-8. Guide for Insect Control.

Type	Habits	How to Eliminate
Ants (carpenter)		
Black, ½ to ¾ inch long. Common along the eastern and midwestern United States.	Usually nest in trees but could do so in the structural areas of your building if the conditions are conducive, such as decaying wood. They do not eat wood but excavate for nesting spaces. They cause considerable damage to wood framing.	Look for sawdust droppings or listen for rustling sounds. Spray nest or ant hills with diazinon.
Ants (house)		
Black, brown or red in color. Live in colonies.	Come into your building looking for greasy or sweet food. Typical entrances are windows, pipe entrances, and any openings or cracks such as around doors.	Treat outdoor ant hills with a residual dusting of diazinon granules and treat entrance areas with diazinon spray.
Bedbugs		
The mature bedbug is brown, oval in shape, and wingless. Its typical size is ¼ to ⅜ inch.	These insects are active only at night but can feed during day if the light is dim. Usually live close to sleeping quarters where they attack and feed on humans. Usually enter homes by traveling on clothing, laundry, secondhand beds and bedding, and furniture.	Malathion or pyrethrum sprays are effective against bedbugs. Treat hiding places such as baseboards, cracks in walls and floors, and furniture. Only treat mattress if label on spray says it is suitable.

Table 9-8. Continued.

Type	Habits	How to Eliminate
Beetles (carpet)		
Winged black or variegated in color. The larvae are covered with golden brown hairs.	The larvae eat woolens, furs, carpets, clothes, upholstery, and other larvae live in dark undisturbed areas where they feed.	To eliminate, shake out, brush out, and air infested clothes and blankets. Dry cleaning can help kill the remaining larvae. Treat rooms with diazinon, malathion, or methoxychor according to label directions.
Beetles (powder post)		
Small, brown, and usually less than a ¼ inch long in size.	Found in wood flooring, structural timbers, cabinets, furniture, and other wood items. Beetles bore round holes approximately ¹⁄₁₆ to ⅜ inch in size. Wood powder or pellets can be found under infested wood. They usually stay away from painted or finished surfaces.	You can use deodorized kerosene to treat wood to control them. Badly infested wood should be replaced.
Cockroaches		
A broad six-legged noctural insect. Brown in color and ranging from ½ to 1 inch in size.	Cockroaches infest the kitchen, bathroom, and other areas of buildings. They spread diseases by contaminating food with their infected droppings. Wet house plants, the bottom of appliances, wet clothes, and other dark moist areas can be a prime nesting site.	Sanitation is the first line of defense. Clean up food scraps including crumbs after every meal. Keep food boxes sealed tightly and avoid leaving dirty dishes in the sink. Diazinon, malathion as well as baits laced with boric acid are effective control measures.
Fleas		
Very small, wingless, dark brown insects.	Fleas respond to anything that moves and to body	Use a good flea powder, spray, or dip and also flea

Table 9-8. Continued.

Type	Habits	How to Eliminate
	warmth. You can have fleas even if you have no pets. They can be brought in on clothing, and they can show up after a neighbor's pet has paid you a visit. Domestic fleas do not transmit diseases, but they do bite humans and cause allergies in pets.	collars to eliminate fleas from pets. You can also use bendiocarb, chlorpyrifos, or malathion. If fleas persist, call in an exterminator.
Flies		
Four-winged insects suspected of carrying and spreading disease.	Breed in large numbers; feeds on garbage, manure, grass and weed clippings, and decaying organic matter.	Good sanitary practices such as keeping screen doors and windows in good repair will help keep down their numbers.
Mosquitoes		
Annoying biting insects that suck blood from humans.	Anyplace that you have still water can be a breeding place for these insects.	"No pest" strips, approved insecticides, sprays and the judicious use of a swatter can all be helpful ways of controlling them. Prevent breeding by draining all stagnant water from around your property.
Moths (clothes)		
Yellowish-tan color. Wingspan approximately ½ inch. White larvae.	Larvae feed on clothing, blankets, carpets, and all other items containing wool or animal fabric. Damage usually goes unnoticed until too late.	Control with napthalene flakes or paradichlorbenzene crystals. Treat infested areas with an approved spray such as diazinon.
Pantry moths and beetles		
A variety of grain-, meal-,	Beetles and larvae of moths	Discard infested food and

Table 9-8. Continued.

Type	Habits	How to Eliminate
and flour-eating insects make up this population.	infest grain products, dried fruits, nuts, and powdered milk. Usually brought into a building with packages of food.	clean shelves and food bins. Remove remaining food and treat with a listed chemical spray such as malathion or diazinon. Keep all food in tightly sealed jars.
Silverfish		
Wingless insects with long feelers and bristly tails.	They damage paper products, books, and wallpaper. Usually active at night and hide during the daytime.	Treat cracks and crevices with diazinon, malathion, or propoxur.
Termites		
White, social, wingless insects. Reproductives are winged insects.	Termites destroy wood in buildings as well as in nature. Wood near or in the ground is most susceptible. At times they use mud tunnels to get to their destination.	Have an exterminator inspect and treat with an approved termiticide.
Wasps and Hornets		
Large winged insects that deliver painful stings.	Nest in cavity walls of buildings or under eaves of the building overhang.	Best to exterminate at night to avoid attack. Treat with pressurized cannisters of propoxur or malathion insecticides.

Insulation

Whatever money that you spend on insulation will come back to you in energy savings and comfort for both heating and cooling. Insulate your heating pipes or ducts, hot-water pipes and hot-water tank, attic and crawl spaces, and water pipes that are in unheated locations. Your local lumberyard or supply outlet has free literature on the different kinds of insulation that also includes where and how to install it. Be sure to avail yourself of these materials and if your building needs insulation . . . get cracking.

Interior (Building)

If you follow Table 9-9 conscientiously, you should be able to keep the interior of your home in very good condition. Don't try to do it all in one day because you simply can't; rather stick to the maintenance schedules that are based on the months, on the seasons, or ones that fit into your personal schedule.

Table 9-9. Interior Maintenance Chart.

✔ Have heating and cooling systems professionally serviced annually.
✔ Test all electrical disconnects annually.
✔ Test ground fault outlets monthly.
✔ Have fireplace and wood stoves cleaned and inspected annually.
✔ Clean and maintain appliances according to manufacturer's recommendations.
✔ Avoid overloading private sewer systems with garbage disposal discharge.
✔ Use a dehumidifier to control condensation.
✔ Clean or replace filters on warm air systems.
✔ Oil and lubricate motors and pumps.
✔ Keep radiators free from accumulated dirt.
✔ Vacuum warm air registers.
✔ Inspect fan belts for wear and tear.
✔ Clean stove hood vents and filters.
✔ Maintain flooring according to manufacturer's recommendations.
✔ Make sure stairways are safe to use.
✔ Test smoke detectors.
✔ Clean and test sump pump.
✔ Drain a bucket of water from the hot-water tank every six months.
✔ Manually test the relief valve on the hot-water tank.
✔ Insulate heat pipes and ducts.
✔ Grout and caulk tile joints.
✔ Monitor the underside of the roof for leaks (during heavy rains).
✔ Make sure that the attic has year round ventilation.
✔ Be sure to label the main shutoffs for water, heat, and electricity.
✔ Remove mildew from walls and ceilings.
✔ Avoid abrasive cleansers to clean fixtures.
✔ Don't pour grease down drains.
✔ Keep tub and sinks free of hairs.
✔ Keep windows clean.
✔ Vent bathroom and kitchen moisture to the outside.
✔ Avoid using extension cords as a means of permanent wiring.
✔ Don't overfuse your electrical service panel.
✔ Tie a string or ribbon to metal pull chains on light fixtures.
✔ Have security system professionally inspected.
✔ Replace missing outlet and switch plate covers.
✔ Repair minor leaks before they become major ones.
✔ Repair or replace defective or damaged drains.
✔ Add outlets to rooms with insufficient ones.
✔ Lubricate locks and hinges.
✔ Clean and service humidifiers.

Table 9-10. Clearances for Wood Stoves.

✓ Stove pipe—minimum of 18 inches from combustible materials.
✓ Wood stove—minimum of 36 inches from combustible materials.
✓ Chimney height—minimum of 3 feet from roofline.
✓ Chimney height must be a minimum of 2 feet higher than any part of the roof within 10 feet, measured horizontally.
✓ An interior masonry chimney must have a 2-inch clearance from all wood framing.
✓ An exterior masonry chimney can have a 1-inch clearance from wood framing because exterior chimneys operate at cooler temperatures.
✓ Noncombustible floor bases must extend not less than 18 inches on all sides of the heating appliance.
✓ Reduced clearances are possible with some kinds of clearance reduction systems and materials. Be sure you check with your local building or fire inspection departments before you make any clearance changes.

Junction Boxes

Open junction boxes are an invitation to an electrical jolt. If you find junction boxes without covers anywhere in the building (basement, attic, crawl spaces) be sure to provide a metal cover for protection.

Kitchens

Your kitchen is probably one of the most expensive areas in your home. Just think about all of the expensive appliances that you have there in addition to the costly cabinets. With this in mind, you should make an effort to save and store in a safe place all of your warranties and instructions for operating these appliances as well as any maintenance materials that you might have. Develop some sort of recordkeeping (refer to Tables 9-12 and 9-13) to help you when it comes time for maintenance and when you are considering replacements.

Kitchen Stoves

Yes, elbow grease is what you need for your stove even though it might have a self-cleaning oven. One fairly simple way to clean it is by putting a bowl of ammonia in the oven overnight. The next day you should be able to wipe off the grease and grime. Be careful though, ammonia could ignite in gas ovens; be sure to turn off the pilot light before you put in the ammonia. Clean the outside of your stove with a nonabrasive cleaner such as baking soda sprinkled on a damp cloth or sponge. If you have a gas stove, clean and dust around the pilot lights and air vents on a monthly basis. Reflector pans on an electric stove can get pretty grimy within a short period of time; clean these off as often as they need it. If your stove has a hood, clean the filter once a month by scrubbing it in sudsy water or by running it through the dishwasher. Dry it thoroughly before you replace it.

Table 9-11. Landscape Maintenance Guide.

Fall

  ✓ Rake fallen leaves every several days to keep them from accumulating.
  ✓ Use fallen leaves as a mulch under shrubs and trees.
  ✓ Shred up larger leaves, such as maple, and put into compost heap.
  ✓ Clean up flower and vegetable beds after they are done bearing.
  ✓ Till flower and vegetable beds and mulch over.
  ✓ Feed cool-season lawns.
  ✓ Make repairs to fencing.
  ✓ Patch damaged sections of the lawn.
  ✓ Start putting away lawn ornamentals and decorations.
  ✓ Support and stake young saplings.
  ✓ Protect tender plants from winter damage.

Winter

  ✓ Start planning for next year's vegetable and flower beds.
  ✓ Order seed catalogs.
  ✓ Make a list of seeds and plants to be ordered in the spring.
  ✓ Prune roses, trees, and shrubs before snow falls.
  ✓ Strengthen all posts and rails on arbors to avoid snow and ice damage.
  ✓ Repair and maintain lawn and garden equipment and tools.

Spring

  ✓ Fertilize lawns and plants as soon as vigorous growth occurs.
  ✓ Start watering as soon as dry spells occur.
  ✓ Mulch to control weeds.
  ✓ Start shearing hedges when they begin to look ragged.
  ✓ Check weekly for insect and disease problems; apply controls if needed.
  ✓ Kill weeds as they start to emerge.

Summer

  ✓ Continue your weed control program.
  ✓ Water plants long enough to soak the soil down to the roots.
  ✓ Remove dead flowers to promote additional flowering.
  ✓ Prune trees where necessary.
  ✓ Fertilize every six weeks during the growing season.
  ✓ Provide shade for heat-sensitive plants.
  ✓ Plan ahead for fall and winter gardens in warmer climates.
  ✓ Continue your garden pest control program.
  ✓ Enjoy the bounty of your garden and the beauty of your flowers.

Table 9-12. Record of Building Maintenance.

Description of Work Done	Work Done By	Materials Used	Date	Costs
Clean gutters	Tony S. Daniels	Linseed oil	9/6/86	$35
Carpenter ant treatment	Matilda Chemical	Diazinon	9/19/86	$125
Trim up trees	Stacy Tree Co.	(labor)	9/24/86	$326
Exterior painting	Steve Bruno	Chelsea Green	10/8/86	$775
Burner tune-up	Bender Heating	new filter	11/5/86	$75
Septic tank pump-out	Grandpa George Septic Service	(labor)	11/7/86	$110
Chimney flues cleaned	Jerry Foster Chimney Sweeps	(labor)	11/17/86	$95
Replace faucet	Don Rudder Plumbing and Heating	Kohler faucet	12/3/87	$119
Clean apartment	Maria's Merry Maids	(labor)	12/8/87	$475
Replace front door	Mario (Mike) Fagone Carpenter	(stock and labor)	12/18/87	$650

Linseed Oil

Linseed oil is a lubricating oil that serves many useful purposes. Mixed with an equal amount of turpentine, it can be used as a wood gutter preservative. Small amounts added to oil-based paints gives the paint more body and better adhesion, particularly to new wood. You can also use it to rub down wood handles for longer life as well as oiling metal parts of tools to prevent rusting. Add some color pigment to it and you end up with a homemade stain. Always keep some of this truly useful product in your building.

Mildew

Mildew is a fungus that grows on moist surfaces. To keep it from becoming a tenant at will, use a dehumidifier or open up windows to ventilate rooms. Use exhaust fans as well to air out moist areas. To remove and destroy mildew, wash the infected areas, which show up as a black or greenish color, with a mixture of one part liquid bleach and three parts water. You will also find mildew and its first cousin, algae, on exterior walls and surfaces. Again, treat with bleach and water and provide more sunlight by judiciously trimming back vegetation that is too close to the building.

Table 9-13. Record of Equipment Repairs and Replacements.

Equipment	Date Purchased	Warranty Period	Life Expectancy	Date	Repairs By	Problem	Cost
Hot-water tank	1/19/75	5 years	5-10 years	2/22/86	Ace Heating	Replace	$450
Window air conditioner	5/4/80	5 years	10 years	4/16/86	Zunke Appliances	Replace	$380
Washer	2/6/87	3 years	10 years	2/22/87	Zunke Appliances	Faulty hose	No charge
Stove	7/2/81	1 year	15 years	8/9/86	Zunke Appliances	Defective filament	$98
Oil Burner	2/8/72	1 year	10 years	10/7/86	Ace Heating	Replace	$600
Central air conditioner air compressor	7/18/71	7 years	12 years	6/10/87	Matilda Air Conditioning and Heating Co.	Replace	$950

Microwave Ovens

Because it is a very good idea to test your microwave oven for radiation leaks, splurge and buy yourself a microwave leakage tester. These can be purchased at most appliance stores. For general safety, follow these guidelines:

- Always follow the manufacturer's directions for clearances. If your unit sits under a shelf, allow at least 1 inch clearance.
- Have damaged door locks and gaskets repaired immediately by a professional.
- Never use metal of any kind in your microwave oven as it could cause electrical arcing and damage the interior.
- *Never* try to repair a microwave oven yourself. Have a professionally trained serviceman do that kind of repair.

National Electric Code

Any and all the questions that you might have as to how a building must be wired are answered by the National Electric Code. Check with your local library for a copy or write to:

National Electric Code
National Fire Protection Association
1800 N. Street, N.W. Suite 570
Washington, DC 20007
(202) 466-3650

Outlets

Don't tolerate outlets that have no covers and/or are loosely mounted on walls. Consider updating older two-slot outlets with the more modern three-slot type. Consider buying yourself an inexpensive outlet analyzer to check outlets for proper grounding and correct polarity. For about 10 dollars or so, you can get a tester from almost any electrical supply store.

Painting

Once a year, go around both the exterior and interior areas of your building and touch up chipped and peeling surfaces. Once you have prepared for painting by repairing minor cracks with an appropriate spackling compound or wood putty, half the battle is done. Also, be sure to reset popped nails in siding or walls and putty over these nailhead holes as well. Be sure to prime the areas that are to be painted with a good quality oil-based primer. If surfaces are rough to the touch, sand them thoroughly before you prime them.

Plumbing

In cold climates, just before winter sets in, be sure to close shutoff valves to the outside so they won't freeze up. If you close up your building for the winter, also drain the plumbing system of its water. Use a nontoxic antifreeze in all sink and toilet drains as well as tubs. Periodically check areas of corrosion on pipes for evidence of active leaks. Insulate pipes that sweat to curb moisture problems in your building. Lastly, label all shutoffs so you can tell which fixtures and appliances are controlled by which shutoffs.

Radiators

At the beginning of the heating system and during it, bleed air from forced hot-water radiators. If you have a steam system, check the air valves to make sure that they are not clogged up with dirt or paint. Replace defective valves rather than repair them because the cost is low enough and the effort usually isn't worth making the repairs. The control valve that feeds steam into the radiator should always be fully on or completely closed. If it is partially closed, you might experience leaks at the valve. Never block the circulation of air around either hot-water or steam radiators as this interferes with their efficient performance.

Refrigerators

Every few months, vacuum the condenser coils in the back of your refrigerator with a brush that you attach to your vacuum cleaner. Once a year, test the door gasket seal by closing the door on a crisp, new dollar bill. Do this several times at different spots. If you can put it out easily, every time, you should replace the gasket. You can help extend the life of a gasket by rubbing mineral oil on it every 6 months.

Roofs

Snow, ice, and wind all take their toll on a roof. Check after each storm and every spring for evidence of damage. Have repairs and replacements made if you see torn shingles, spots where shingles are missing, cracked and damaged tiles, or rotting wood shingles. Always avoid walking on roofs for your personal safety as well as to avoid damaging shingles. Because working on roofs is a particularly hazardous activity, it would be wise to hire a roofer to do this work for you.

Rust

Don't let rust get a head start on any metal portions of your building. When you find it, be sure to sand down to the bare metal, prime with a quality metal primer, and apply two coats of a paint formulated for use on metals.

Septic Systems

Every year inspect your tank and field for signs of problems. Have the tank pumped out every three to five years. Do not discharge grease, paints, derivatives of lye, sanitary napkins, colored or printed toilet paper, garbage, or trash into your system. If your garbage disposal is directly hooked up to your septic system or cesspool, it might be a good idea to remove it. Disposals create an added burden on the tanks and fields. So you won't overload your tank with excess water, have a separate dry well installed for your washing machine. Never build anything over the tank or the field as this could cause some serious damage and would make access to the tank difficult or impossible. The cover to the tank should be readily accessible at all times for inspection and cleaning purposes.

Smoke Detectors

Every few weeks press the test button on each of your smoke detectors, which should then give off an audible alarm; if they don't, replace them. Every 3 months or so, test them by blowing smoke from a match toward them. Again, each one should go off. To stop them, blow fresh air in their direction. Always replace batteries at the time the built-in indicator signals a low-current warning. If you don't happen to have smoke detectors in your building, have them installed because in addition to being foolish you might also be in violation of local codes.

Steam Heating System

Every 2 to 4 weeks during the heating season, blow off the low-water cutoff valve to drain sludge from the bottom of the chamber. Failure to do this could result in a defective safety valve. Each time you are in the basement, "eyeball" the sight glass to make sure that there is sufficient amounts of water in the gauge. Usually half to two-thirds full is fine. Also, now and then check the boiler's pressure by viewing the pressure gauge. As is true with most domestic steam systems, your pressure gauge should not read above a few pounds pressure. If you find a reading of several pounds pressure, shut off the system and call in your serviceman.

Storm Windows

Sometime during the spring and again during the fall, check the weep holes at the base of the storm windows to make sure that they are still there. These allow trapped moisture to escape before it condenses to water and becomes troublesome. In the event that you have no weep holes, simply drill them in at each end of the window sills.

Sump Pump

Before the flooding season starts in your area, check the sump pump screen at the base to make sure that it is free of debris and is generally clean. To check the switch operation, submerge the pump in a bucket of water. If the pump is functional, it will kick over and pump the water out of the bucket. Carry out this procedure every month so you will know the pump will do its job when needed.

Thermostats

Once a year, remove the cover of your thermostat(s) and dust very carefully with a light, soft brush. Clean the metal contacts by sliding a clean card between them several times. Be sure to never block the free circulation of air near your thermostat. Furniture, drapes, and the like have no place in front of it.

Trash Compactors

So you don't have to clean constantly and forever, wrap your wet garbage in newspapers and avoid putting anything that has a strong odor into your compactor. Follow the manufacturer's directions as how to remove the ram so you can clean it in hot, soapy water. To get rid of odors, spray with a disinfectant spray. Eliminate squeeking and screeching noises by lubricating and adjusting the drive belt.

Untreated Wood

Decks, porches, and any wood that is exposed to the elements should, where practical, be built with treated wood. This holds true for those parts of a building that are close to or in direct contact with the soil, such as steps and foundation windows. Untreated wood, even though it might be well stained or painted, invites decay and damage by wood-boring insects. For the few extra dollars that the pressure-treated wood costs, you will reap the benefits of never having to paint or stain again nor will you have to replace rotting wood and wood-boring insect damaged wood.

Ventilation

To keep your building virtually free from humidity and condensation problems, there is only one thing to do—and it has been said often enough—ventilate, ventilate, and ventilate. If you are troubled with condensation, mildew, musty odors, and sweating pipes, open up the building by adding vents to attics, crawl spaces, exterior walls, and anywhere moisture could get trapped.

Warm-Air Heating System

There are several easy maintenance jobs that you can do to make your hot air system function efficiently. During the heating season—on a monthly basis—clean or replace the air filters (they cost as little as 75 cents). A filter that is dirty could cost you an additional

10 to 20 percent monthly in heating or cooling costs. Tests have shown that dirty filters can actually contribute toward the destruction of an air-conditioning compressor, a heat pump, or a heat exchanger in a gas- or oil-fired furnace. So get with it—get rid of dirty filters!

If you have an electronic air filter, remove it from the furnace and clean it with mild detergent and water every two or three months. Lubricate all oil cups on motors and blowers according to the manufacturer's guidelines. Check fan belts for deterioration and adjust or repair them as needed. And don't forget to vacuum your floor, wall, or ceiling registers regularly.

Washing Machines

Washing machines can suffer from worn or kinked hoses and used up washers. If you don't take care of those things, you might need to do some mopping up shortly from an unexpected flooding from the machine. Make sure that the surface that the machine is on is solid and level so it will not vibrate itself to death. As part of your maintenance program, be sure to wipe frequently around the rim and empty out the lint trap on a regular basis.

Wells

Once a year, or at the time when your water changes color and starts to taste funny, have it tested and analyzed for bacterial contamination. Put a sample of water in a sterilized bottle and take it to your local or state public health lab. Draw your sample in the morning when the concentrations of contamination is highest.

Whole-House Fan

At the beginning of the cooling season, check the louvers of your fan for leaves and debris that could block them. Also, lubricate according to your owner's manual and check the fan belt's tension. If there is more than ½-inch deflection when you press it gently, have it tighten or even replaced if it shows wear. At the end of the cooling season, seal it with an air-tight cover and put insulation over it to keep out any winter drafts.

Windows

Lubricate window tracks with beeswax to give them some help in going up and down. Clean window panes with a solution of one cup vinegar to 1 gallon warm water. For very dirty ones, use a solution of 1 tablespoon household ammonia, 3 tablespoons vinegar, and 1 quart warm water. Don't throw away your old newspapers, instead use them to dry and polish your window glass. Every spring and fall check the putty and caulking on both the storm and on the house windows. Be sure to replace missing portions.

Wood Stoves

Just before the heating season, double-check the clearances between your wood stove and all combustible materials. There should be nothing so near as to constitute a fire hazard. Table 9-10 lists some of the recommended clearances for wood-burning stoves. In addition, see to it that the stove, the stovepipe, and the chimney flue liners are clean and free of creosote buildup. If they need it, always repair the chimney and stove before using them.

Yard Work

No matter if you are an avid gardener or just a weekend landscape warrior, keep in mind that there are timetables for the yard. Table 9-11 gives you a rough idea what the four seasons of landscape maintenance entail. Like your other maintenance jobs, make the work fit into your specific needs and desires. If all you must do is cut the grass, then you really won't need much of a schedule. But if you fall into the category of the dedicated yard person, you should put yourself on a more structured schedule. Table 9-11 is just a sample; be sure to change it to fit exactly what you need.

Zinc Chromate

Zinc chromate is a bright yellow pigment and is used in paints as a rust-inhibiting agent. You can buy it at any good paint supply store. It can be used directly from the can as a primer for metal. It is particularly useful for exterior railings and fences. If you want a job that will last, sand down the rusted areas to bare shiny metal and prime them with zinc chromate before you paint.

CONCLUSION

Just to remind you again, always keep a record of your maintenance work as well as of repairs that you have done or have had done for you. Tables 9-12 and 9-13 give you some idea as to how to set up the whole scheme. Be sure to describe the work done and jot down the date and the name of the contractor that did the work. Information about warranties and anticipated life expectancies give you a chance to compare and see what you can expect from specific equipment. By keeping records such as these, you will be able to check back to see if you did the job that you were supposed to have done and if it is time to repeat some maintenance chore. Leaving it all to memory is just asking a bit too much. You might just find yourself doing what you want to do and forgetting about some more tiresome chores. Not a very good idea.

Please keep in mind that this chapter covers a lot of territory and that it is only meant as a quick overview, certainly not as an in-depth report. Always refer to manufacturers' recommendations and suggestions in their manuals for specific information on what to do, when to do it, and how to do it.

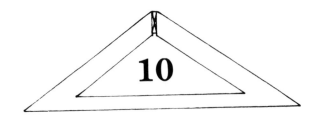

How to Select a
Contractor

Although this book is essentially about do-it-yourself building maintenance, Chapter 10 is included for those times when you need the services of a professional contractor for repair or renovation work too complicated or time-consuming for you. A contractor has the necessary expertise, equipment, and tools to do the job in half the time it might take you to do it. And who wouldn't want to have some spare time.

Before you call a contractor, however, there are a number of considerations that you should take into account. To begin with, you have to decide if the work to be done and the money you plan to spend on it are really worth it. Clearly, if your heating system or septic tank needs repairs or replacement, there should be no second thought in your mind about having the work done. If, on the other hand, you are contemplating a major renovation, such as putting in a new bathroom or adding a greenhouse to the side of your home, you should seriously plan ahead. So, before discussing the guidelines for selecting a contractor, let's consider other variables that will affect a project.

LIMITING FACTORS

One consideration that you should take into account before you make elaborate plans for remodeling and/or renovations is the selling price of buildings or homes in and around your neighborhood. Let's say that buildings in a given area sell for $150,000 to $200,000, and your renovation marks up the value of your property to $300,000. You might very well end up with a "white elephant"—a difficult property to sell. In fact, the chances are that you will only recover approximately 20 percent of the money that you put into it.

Deed restrictions sometimes prevent owners from making changes or additions to a building. Check your deed to make sure that there are no such covenants for your property.

Also, local zoning laws determine how land and buildings can be used in a given community. Such ordinances regulate land usage; front, side, and rear setbacks; and the height of fences and hedges. If you have any questions, call your local building department to clear the improvement plans that you have. If zoning ordinances affect your proposal, you have the right to petition for a variance (an exception to the law). You might need a lawyer for this, so check with local officials to find out how to apply for such a variance.

Building codes set minimum standards for the construction and remodeling of buildings and act as limiting factors. The contractor that you hire is responsible for complying with these codes. If you hire a contractor that is either unlicensed or does not follow local codes, you might be in for some real headaches. Get a copy of the local building code at your town or city hall. It is not only very handy to have, but it might also prove to be interesting reading.

RENOVATIONS THAT PAY OFF

Once you get past all the limiting factors and you are sure that your plan can get off the ground, you will still want to make sure that the investment of time and money is worthwile. If the renovation makes your home more attractive, more liveable for you and your family, and if it enhances its value overall, then you are indeed heading in the right direction. Table 10-1 lists the types of improvements that pay off and are worth doing. Kitchen remodeling and bathroom additions usually are the kind of improvement that will increase the overall value of your property. Also high on the list are central air conditioning or adding a garage because they each will add to the general well-being of your existence. In short, if you take your time and choose your improvement well, you won't have to wait for resale to profit from your investment but will reap the benefits from the day of its completion.

ADDITIONS THAT DON'T ADD UP

If payback is important to you, swimming pools, tennis courts, and elaborate landscaping won't help you much. Most buyers don't want to get involved with a lot of extra maintenance chores. Putting in a $30,000 swimming pool into a $100,000 property, for example, is definitely out of proportion, and the payback will be many lap years away. Putting up expensive fence or screening won't cover up the fact that you live in a not so exclusive area, so don't waste your time or money. Expensive shrubs, fancy statues, and ''golly gee'' plantings might bring some oohs and aahs but very little in the way of big paybacks. Be sure to consult Table 10-2 before you start any project that might be questionable.

Don't overbuild is one of the key guidelines you should follow in your renovation program. Keep the costs of the work under 30 percent of your building's current market value. This is especially true if, once the work is completed, your property is worth far more than the average-priced building in the neighborhood. Keep in mind that people who are willing to pay top dollar usually also want an exclusive neighborhood where each home falls more or less into the same price range.

262

Table 10-1. Improvements that Pay Off.

IMPROVEMENT	APROXIMATE COST	POTENTIAL RETURN
Kitchen remodeling	$2,500 to $25,000	Probably one of the best returns. Select energy-efficient appliances.
Addition or remodeling a bathroom	$1,000 to $10,000	Smart investment with a high return.
Addition of a second story	$20,000 to $35,000	Ideal for young families with growing pains.
Addition of a family room	$10,000 to $30,000	Extra first-floor living space makes good sense. Below grade rec rooms have less appeal.
Functional space in the basement (workshop, laundry or exercise room)	$1,500 to $5,000	Worth the effort. Makes the building more attractive to a buyer.
Central air conditioners	$2,500 to $7,500	Usually worthwhile in warmer climates. Less payback in the northern states.
Energy-efficient fireplace or solid fuel stove.	$500 to $5,000	Usually a good payback. Location is important for efficiency and pleasing arrangement of furniture.
Wood deck or porch	$2,500 to $7,500	Usual return is between 30 to 60 percent of the initial cost.
Two-car garage	$10,000 to $25,000	Very good investment and a big selling point if you decide to sell your property.
Attached solar greenhouse	$10,000 to $30,000	Gaining favor in higher priced homes.

KEEP IT SENSIBLE

Whether you are building a new addition or updating an old bathroom, be sure to make changes that can appeal to others as well. Make changes that are in proportion with the rest of the building. Neutral colors, especially in permanent fixtures and appliances, are preferred by most people, to name one simple example.

Table 10-2. Additions that Don't Add Up.

ADDITION	POTENTIAL RETURN
Elaborate landscaping	You will be lucky to get back 25 percent of your investment when you sell. Keep plantings in tune with the size and price of the building.
Swimming pool	Poor investment other than for the pleasure you get from it. Most people don't want the maintenance and risks involved.
Basement rec rooms	You will be lucky to get 30 to 40 percent return on your investment.
Expensive screening and fencing	Fair to poor return for the money spent.
Tennis court	Very poor return unless the buyer is into tennis.
Detached greenhouse	Poor return unless the buyer is an avid gardener.
Energy-saving improvements	Add little to the sale of a property because buyers have come to take them for granted.
Gazebos, cabanas, shuffleboard courts, dog kennels and tree houses	Forget it; you will never recoup these expensive frills.
Central vac systems	Usually add to the value only in more expensive buildings.
Intercom system	You might start talking to yourself when you see how low the return on this is. How about 10 percent!
Solar heat or hot water	The return generally is not good, particularly in the northern states where backup systems are required.

It is particularly important to take the style of your building into account if you plan on making substantial changes to the outside. Just imagine an adobe-styled addition to a New England-styled cape! That would be quite an eye-catcher, but that is about all.

Above all, be sure to have a licensed professional, who follows local codes and ordinances, do the work. When you do hire a contractor, go over the checklist in Table

Table 10-3. How to Select a Good Contractor.

✓ Check with friends, neighbors, and relatives for referrals.
✓ Ask at your local building department for the names of contractors.
✓ Professionals such as architects and real estate brokers might be of help.
✓ Call your local home builders association for names of qualified people.
✓ Double-check with the Better Business Bureau and the Consumer Protection Council.
✓ Follow up with calls to contractor's clients.
✓ Be sure to get more than one or two estimates.
✓ Stay away from a very high or very low bids.
✓ Find out if the contractor has a legitimate business address and how long he has been at that location.
✓ Check if the contractor is licensed and bonded.
✓ Get a list of recent clients and try to see the finished work.
✓ See if the contractor has done similar jobs to the one you have in mind.
✓ Find out how long the contractor has been in business.
✓ Do you feel comfortable talking to the contractor?
✓ Can you talk to the contractor and does he listen to what you say?
✓ Do you have a good "gut reaction" from your dealings with the contractor?
✓ Is the contractor insured and does he carry workman's compensation?
✓ Does the contractor offer a warranty program?
✓ Would the contractor mind having a professional monitor his work?

10-3. If you don't like what you are hearing from the contractor, move on to another interview with a different contractor. Although you might like the lowest price bid, if the workmanship is going to be below standard, it will cost you twice as much in the long run. In fact, inferior workmanship can lower the value of your property considerably.

SELECTING THE RIGHT CONTRACTOR

With your plans in hand and with the knowledge that your improvement will bring joy to you and your family, you must now embark on the quest for the right contractor. It won't be easy, but it will be worth all of the effort that you put into it.

How do you get started on your search, though? First, ask people who have had work done recently—coworkers, friends, relatives, and neighbors. If you are unable to get names that way, try the local building department. People there should know which contractor is reliable and which is not. Often building supply stores have a list of reputable contractors and can recommend one or two in particular. Architects, bankers, and lawyers could probably also be of help. An excellent source for names of qualified contractors is your local Home Builders Association.

After you have compiled a list of names, set out to verify their track records. Check with the local Better Business Bureau or Consumer Protection Council to find out if outstanding complaints were issued against contractors that you are considering. Also check with the local Building Department to see if their inspectors have had a hard time making these contractors stay within the building codes. Names of customers that contractors gave you as references should be contacted. Sometimes one call can make the difference between signing on the dotted line or bowing out of the deal.

After you have narrowed your list down to three or four names of professionals that you feel comfortable with, the bidding begins. Each contractor will examine and study your draft and proposals of what you want done. It is crucial, though, that you know exactly what types of materials and products you want—for example, what kind of tile or what model and brand of an appliance—otherwise the difference between one bid and another will be so great that you won't know which to believe. The same specifications must be given to each bidding contractor so that the results will be fair to all involved.

When all bids are in, look each one over carefully and make sure that all specifics are listed, including quantities and grades of materials to be used. Most reputable contractors provide construction drawings as well to give you an idea as to what your bathroom, for example, will look like after the work is completed. By the way, estimates are usually given without cost and with no strings attached.

Which bid should you accept? First, be wary of the lowest bid, especially if it is considerably lower than all the others. Maybe the contractor omitted something or he might be planning on using inferior materials. A strikingly low bid might also mean that the contractor is not realistic as to what the work will entail so that he will have to cut corners later on. Eliminate him. Because the other estimates will still vary somewhat, you will have to use additional yardstick measures to zero in on the one contractor that you will want to select.

With the lowest bidder eliminated, you still have to decide about the other three or so. If one of the final bidders is approximately 10 percent higher than the others, eliminate that one as well. Coming down to the wire with two close bids means that you still have to put on your detective cap and do some investigating. Find out if the finalists have legitimate business addresses and phone numbers. Ask them how long they have been in the business and if they have had experiences with and are licensed to do the type of work that you want to have done. Are they bonded and do they offer warranties? What type of insurance do they carry? Are the workers covered by workman's compensation and does the contractor carry general liability insurance? If a ladder falls and damages a neighbor's house in the process, who will pay the repair costs? You should get answers to all of these questions. In addition, they should supply you with three recent customers whom you will call to check their work. You could perhaps even take a look at some of the work these contractors have done to get an idea as to the quality of the workmanship. If you like what you see, ask yourself how well you get along with them—put some weight on your ''gut'' feeling towards the finalists. And, at last, the one you feel most comfortable with discussing the job is the person that you will want to hire.

WARRANTY PROGRAM

If you really want to go the safe route because the job you need to have done is very substantial, you might want to look for a contractor who participates in a warranty program, such as the HOW Plan. Many contractors participate in this warranty program that provides for a 5 year warranty plan of major structural and mechanical systems. In addition, the plan also provides an informal dispute resolution program between a contractor and a building owner if any disagreements arise. Keep in mind, though, that you don't get anything for nothing; and you will have to pay for this program, if not directly, then in a higher contract price. If you are interested in detailed information about this program, write to:

Home Owners Warranty Corporation
2000 L Street, N.W.
Washington, DC 20036
(202) 463-4600

Two useful pamphlets that could further help you in selecting a contractor are entitled: *Remodeling Without Worry* and *A Homeowner's Guide to Selecting a Roofing Contractor*. The first is available from the above address. The second pamphlet can be obtained by writing to:

Bird Incorporated
Department CD
Washington Street
East Walpole, MA 02032

CONTRACTS

Now that the job to be done is clearly defined and the quest for the contractor is over (see Table 10-3 for a recap on selecting a contractor), you now have to work on a contract. The bidding estimate and the verbal agreement are simply not enough to protect you. So that every party is on the safe side, it is essential that all clauses are clearly spelled out in a written agreement. As is true with any legal contract, it is prudent to have your attorney examine it before you sign on the dotted line. Make sure that you understand all the terms in the agreement and that you speak up if you have any doubts or questions.

As you can see in Table 10-4, a good contract should include many important features. The name of the contractor, his business address and phone number, as well as his license number should be included. Write the specific work to be done in outline form, though

Table 10-4. What Should Be in a Good Contract.

✓ Name, address, and telephone number of the contractor
✓ License number of the contractor
✓ Detailed description of the proposed work
✓ The cost of the work and the payment schedule
✓ Work not to be done by the contractor
✓ All warranties and guarantees
✓ A three-day cancellation clause
✓ The starting date and the completion date
✓ Penalty clause for late completion
✓ 10% holdback clause
✓ Broom clean clause
✓ Bond clause against liens
✓ Type and amount of insurance

it must be detailed enough to avoid misunderstandings. Many contracts include the phrase, "all work to be done in a workmanship manner," which means that the standards accepted locally will be adhered to. The contract should also include the cost figures for the job as well as how the payments are to be made. Most call for one-third of the cost to be paid up front, one-third at the completion of half of the work, and the remainder at the total completion of the job. On jobs that require a lot of work hours, a clause is written in stating that 10 percent of the price will be held back for 30 days after the completion of the work, which gives the owner sufficient time to inspect for defects.

Most attorneys will also want to see a date as to when the work is to begin and when it is expected to be completed. Another clause that could be included is to penalize the contractor if the work goes on beyond a reasonable time. The phrase "time is of the essence," is a legal way of saying that the work better be completed by a certain time period.

Additional clauses in a contract could include warranties and guaranties that the contractor agrees to and disclaimers that the contractor is not going to be responsible for, such as painting or wallpapering. Other responsibilities on the part of the contractor, such as removal of debris and cleaning up after each work day, should also be included in the contract. Cleaning and the removal of debris is also called the "broom clean clause," which is important because some contractors will—if allowed to—leave your building in shambles until the last day. A well-seasoned attorney might want to include the clause for a contractor to post a bond protecting his client from liens filed by suppliers or by the contractor's employees. The type and amount of insurance coverage that the contractor carries should also be included in the contract.

Finally, include a 3-day cancellation clause. This gives the buyer of the contractor's services three days to think it over before it becomes a final legal document. If during those three days after signing the contract the buyer wants to terminate it, he can merely send a formal letter stating his reasons to end the agreement.

One final note on contracts: like most things in life, adjustments and changes are inevitable. If any changes are made in the contract after work has begun, they must be put in writing and new cost and time factors must be noted as well. All such changes must be initialed by all parties concerned. As with the original contract, let an attorney examine it before any terms are finalized.

INSPECTING THE WORK

It might be an added expense, but it is well worth it to hire a professional building inspector to monitor the work as it progresses, particularly if it is a major renovation job. The inspector would come in and check the work from the start and would continue to do so on a weekly basis. Such checkups keep the contractor and the work crew on their toes and accountable because they know that someone who understands construction work will be coming in periodically to check for sloppy workmanship and poor quality materials. After the work is completed and before you sign off, the inspector will make his last walk-through to see if it is necessary to give the owner a "punch list" of defects that the contractor will have to rectify. Once the inspector is satisfied that the work has been done according to the contract, you would make your last payment to the contractor.

Epilogue

Sad but wiser, you cannot help but look back and say that the acquisition of your property did not put an end to your sweat and tears. If anything, the purchase was just the beginning of a long-term contract with yourself for an ongoing inspection-maintenance program. To keep its worth or, better still, to increase its worth, you must now provide constant care to this, your major investment. Armed with the information in this book, however, you don't have to just sit back and wait for the repairman to come and perhaps walk away with your checkbook. Knowing what to do and when to do it is now at your fingertips. Be sure to review each section of the book and start to develop a routine that you can live with for both the inspections and maintenance of your property. Owning real estate should be both enjoyable and profitable, but it won't be if you don't follow the rules set down in this book.

Many a successful building owner, after learning about the keys to good maintenance, goes on to buy more properties for the purpose of investment or just to own a summer or winter retreat. You can do the same. Maybe you will be one of the lucky ones that can parlay the information found in *What's It Worth* and *Keep Its Worth* into a dream come true. I certainly hope so. Good luck and best wishes for a successful building maintenance program.

Appendices

Appendix A
Inspection/Maintenance Worksheets

EXTERIOR

Area	Date	Condition and Comments
Roof		
Chimney		
Roof drainage		
Exterior walls		
Windows		
Foundation		
Foundation windows		
Entrances		
Ventilation		
Garage/s		
Wood-boring insects		

Driveway

Septic/cesspool

Crawl space

Fence

Trees, shrubs

INTERIOR

Area	**Date**	**Condition and Comments**
Wood-boring insects		
Household insects		
Foundation		
Seepage		
Condensation		
Decay		
Electrical		
Heating		
Air conditioning		
Plumbing		
Hot water		
Attic		
Crawl space		
Insulation		
Bathrooms		
Kitchen		
Fireplace		

Appendix B
Additional Reading

Air-To-Air Heat Exchangers for Housing
William A. Shurcliff
Brick House Publishing
Andover, MA

Appraisal and Rehabilitation of Old Dwellings
U.S. Department of Agriculture
Handbook No. 481

Basic Home Repairs
Sunset Books
Menlo Park, CA

Chimneys, Fireplaces, Vents and Solid Fuel Burning Appliances
Standard No. 211
Fire Protection Association
Batterymarch Plaza, Quincy, MA 02269

Concrete and Masonry
U.S. Department of the Army
Handbook No. TM 5-742

Condensation Problems in Your House
U.S. Department of Agriculture
Handbook No. 373

Corrosion Facts For The Consumer
U.S. Department of Commerce

CPSC Guide To Electrical Safety
U.S. Consumer Product Safety Commission
Washington, DC 20207

Drainage Around Your Home
U.S. Department of Agriculture
Handbook No. 64

Electrical Code for One- and Two-Family Dwellings
Fire Protection Association
Batterymarch Plaza, Quincy, MA 02269

Finding and Keeping a Healthy House
U.S. Department of Agriculture
Handbook No. 1284

Fire Safety in Housing
U.S. Department of Housing and Urban Development
Handbook No. H-2176R

Guide To Roof and Gutter Installation and Repair
McGraw-Hill, Inc.

Home Heating Systems
U.S. Department of Agriculture

Home Maintenance
William Weiss
Bennett Publishing Co.
Peoria, IL

In the Bank or Up the Chimney
U.S. Department of Housing and Urban Development
HUD-PDR-89

Maintenance in the First Degree
Washington Gas Light Company
1100 H. St., N.W.
Washington, DC 20080

Prevention and Control of Decay in Homes
U.S. Department of Agriculture

Principles for Protecting Wood Buildings From Decay
U.S. Department of Agriculture

Residential Asphalt Roofing Manual
Sumner Rider and Associates
355 Lexington Ave.
New York, NY 10017

Roofing Materials
University of Illinois
One East Mary's Road
Champaign, IL 61820

Roofing Simplified
Donald R. Brann
Directions Simplified, Inc.
Briarcliff Manor, NY 10510

Roofs and Siding
Time-Life Books
Alexandria, VA

Safer Products, Safer People
U.S. Consumer Product Safety Commission
Washington, DC 20207

Septic Tank Care
U.S. Department of Health, Education, and Welfare

Simple Plumbing Repairs
U.S. Department of Agriculture
Handbook No. 1034

Solid Fuel Safety Manual
Wood Heating Education and Research Foundation
Suite 700
1101 Connecticut Avenue, N.W.
Washington, DC 20036

The Complete Handbook of Plumbing
Robert E. Morgan
TAB BOOKS Inc.
Blue Ridge Summit, PA 17214

The Energy-Wise Home Buyer
U.S. Department of Housing and Urban Development
HUD-H-2648

The Master Handbook of All Home Heating Systems
Billy L. Price and James T. Price
TAB BOOKS Inc.
Blue Ridge Summit, PA 17214

Ventilating Residences and Their Attics for Energy Conservation
U.S. Department of Commerce

What's It Worth? A Home Inspection and Appraisal Manual
Joseph V. Scaduto
TAB BOOKS Inc.
Blue Ridge Summit, PA 17214

Wood Heat Safety
Jay W. Shelton
Garden Way Publishing
Charlotte, VT 05445

Wood Frame House Construction
U.S. Department of Agriculture
Handbook No. 73

Wood Inhabiting Insects In Houses
U.S. Department of Housing and Urban Development

Government publications can be purchased by writing to:

Superintendent of Documents
U.S. Government Printing Office
Washington, DC 20402

Appendix C
Additional Maintenance
Information Sources

American Society of Home Inspectors (ASHI)
3299 K Street N.W., 7th Floor
Washington, DC 20007

Association of Home Appliance Manufacturers
1901 L St., N.W.
Washington, DC 20036

National Association of Plumbing-Heating-Cooling Contractors
1016 20th St., N.W.
Washington, DC

National Electrical Code (N.E.C.)
National Fire Protection Association (N.F.P.A.)
1800 M St., N.W., Suite 570
Washington, DC

National Insulation Contractors Association
1120 19th St., N.W.
Washington, DC 20036

National Pest Control Association
8100 Oak St.
Dunn Loring, VA

National Roofing Contractors Association
1001 Connecticut Ave., N.W.
Washington, DC 20036

Small Homes Council-Building Research Council
University of Illinois at Urbana-Champaign
One East Saint Mary's Road
Champaign, IL 61820

Solar Energy Institute of America
1110 6th St., N.W.
Washington, DC

Wood Heating Education and Research Foundation
1101 Connecticut Ave., N.W., Suite 700
Washington, DC 20036

Glossary

aluminum siding—Metal siding used on buildings. Buildings with this type of siding material will develop condensation problems unless the building is adequately ventilated.

American Society of Home Inspectors (ASHI)—A national organization of reputable building inspectors. Names of ASHI members can be found in most business telephone directories.

asbestos—A hazardous mineral fiber insulation found in many buildings.

bulkhead—A horizontal or inclined door providing access to a cellar.

burner—That part of a furnace or boiler where combustion takes place.

carpenter ant—Species of ant that nests and burrows in wood.

catch basin—A receptor or reservoir that receives surface water runoff or drainage.

cesspool—A lined excavation in the ground that receives the discharge of a drainage system, especially from sinks, tubs, and water closets.

chemical desiccants—Chemical dehumidifying agents that absorb moisture out of the air.

circuit breaker—An electrical device for discontinuing current to electrical receptacles.

condensation—Moisture that accumulates in a building, frequently on the inside of the exterior portions of a building.

crawl space—A shallow space below a building, generally not paved and often enclosed.

creosote—Chimney and stovepipe deposits originating as condensed wood smoke.

dehumidifier—A device for removing water vapor from the air.

dormer—A structure projecting from a sloped roof, usually with a window.

downspout—A pipe of metal or plastic used for carrying rainwater from roof gutters.

dry well—A pit that is designed to contain drainage water until it can be absorbed into the surrounding soil.

ducts—In heating and air conditioning, the conduit used to distribute the air.

efflorescence—A whitish, powdery deposit of salts brought to the surface of basement walls by moisture.

flashing—Metal or other form of thin, inpervious materials used to prevent water penetrations through building joints and intersecting junctions.

fungi—Microscopic plants that live in damp wood and cause mold, stain, and decay.

fuse—A protective device that guards against overcurrent in an electrical system.

gable—That portion of the end of a building that extends upward to the peak or the ridge of the roof.

ground—The conducting connection between electrical equipment or an electrical circuit and the earth.

ground fault circuit interrupter (GFCI)—A quick-tripping circuit breaker that can cut power within $\frac{1}{40}$ second after detecting current leakage (also called GFIs).

heat pump—A heating and cooling plant best suited to moderate climates. Technically it is a compressor-driven refrigerant cooling system that can function as a heater when the cooling cycle is reversed.

humidifier—A mechanical apparatus that adds moisture to the air or other materials.

ice dam—A formation of ice at the eaves of a roof that allows water to back up and seep into the upper levels of a building.

jumper cable—A short length of cable or wire used to bond the electric grounding connection to both the street side and the house side of the water meter.

lead (paint-pipes)—a health hazard found in lead paint and lead water pipes in older buildings.

ledge—A rock outcropping.

leakage—Water penetrations that flow rapidly, unlike the slower flow of seepage.

low-water cutoff valve—In a steam heating system, the important shutoff valve that turns off the boiler with low water.

maintenance—The act of preventing deterioration and breakdowns in a building's systems.

metal tie-rods—A metal unit used in the construction of concrete foundation.

mildew—A fungus that attacks plants or appears on organic matter, especially when exposed to moisture.

mud tunnel—A shelter tube used by termites to get to wood structures.

muriatic acid—A chemical solution used to clean masonry surfaces.

perimeter drain—A system of buried or surface drains used to carry water away from a building.

point up—To clear loose mortar from brick joints and refill with fresh mortar.

powder-post beetles—Wood-boring insects that turn wood into a powdery residue.

pressure-treated wood—Wood framing that has been infused with chemical preservatives under pressure.

radon—A naturally occurring gas that is caused by the radioactive decay of the element radium.

ridge—The horizontal line formed by the upper edges of two sloping roof surfaces.

seepage—External slow leaks resulting in water in basement areas.

septic tank—A watertight receptacle that receives the discharge from a building plumbing system and is so designed to separate solids from the liquids.

siding—The exterior covering of a building on wall areas.

soffit—The underside of a roof overhang or eave.

stucco—A cement plaster used for finish coatings on exterior surfaces of foundations and buildings.

sump pump—A small pump used to remove water from a sump hole.

termites—Wood-boring insects that resemble ants in size, general appearance, and habit of living in colonies. If left alone, they will eat out sound wood, leaving only a shell to conceal their activities.

urea formaldehyde—A hazardous substance found in insulation, plywood, particleboard, and carpeting.

valley—An internal angle found by the junction of two sloping sides of a roof.

vapor barrier (retarder)—Material used to retard the flow of vapor or moisture into walls and ceilings, thus preventing condensation.

ventilation—A natural or mechanical process by which air moisture is removed or introduced into a building.

vinyl siding—A type of building siding that tends to prevent the movement of moisture out of a building.

weatherstripping—Strips of metal or other materials designed to prevent the passage of air, water, and moisture around doors and windows.

weep hole—A small opening at the bottom of a storm window that allows water to drain out.

Index

288

condensation and, 48, 51, 53
energy conservation and, 134
indoor pollution and, 210-213
inspection of, 6, 9
kitchen and bathroom, 52
leaking around roof-mounted
 units for, 88
proper installation of, 54
proper system for, 53
ridge/soffit, 94
tips for proper, 137
types of vents used for, 136-140
upgrading of, 66, 67, 68
vent pipes in, 87
whole-house fans for, 258
vents (see ventilation), 53

W

walls, exterior, inspection of, 4
warm air heat, 152-154
 maintenance of, 257
 tension belt on fan for, 156
warranty programs, contractors, 266
washing machines, 258
water penetration, 16
 correct repair of, 20
water stains, 82
 siding and trim leakage for, 99
water tables, high, wet basements
 and, 41
waterproofing products, 40
weathering, decay in wood caused
 by, 107
weatherstripping, 131
 application of, 130

reducing air infiltration through,
 129
weep holes, 130
wells, 258
wet basements, 25-46
 buildings on sloped lots and, 41,
  42
 bulkheads causing, 29
 catch basins for, 34
 causes and cures for, 46
 causes of, 26
 condensation and, 65
 dampproofing and, 44
 dry wells for, 44
 floor-wall joint and, 39
 foundation grading for, 27
 foundation plantings and, 31
 foundation problems for, 33-40
 foundation windows causing, 29
 high water tables and, 41
 landscape plantings and, 32
 low spots and, 31
 perimeter drainage system to
  cure, 42
 repointing leaking foundation
  joints for, 35
 rock ledges and, 31
 roof drainage and, 26
 seepage and condensation in, 46
 steps in repair of, 26
 sump pumps and, 43
 surface mounted interior drains
  for, 43
 tree roots causing, 29, 32
 waterproofing products for, 40

whole-house fans, 258
windows
 condensation and, 64
 energy conservation and, 141
 foundation, inspection of, 5
 inspection of, 5, 18, 23
 life expectancy of, 224
 maintenance for, 258
 storm, 256
 weep holes in, 130
winterizing, 182
wood
 soil and vegetation contact with,
  113
 treated (see treated wood), 111
 untreated, 257
wood-boring insects (see also
  insects), 116-126, 245
 carpenter ants as, 122
 carpenter bees, 123
 control guide for, 246
 identification of, 125
 inspection for, 8, 10
 powder-post beetles as, 121
 prevention of, 125
 termites as, 117
woodstoves (see stoves), 17, 18
worksheets, 271-272

Y

yard work, 259

Z

zinc chromate, 259

Edited by Kathleen E. Beiswenger and Joanne Slike